A Guide to Field Instrumentation in Geotechnics

Geotechnical instrumentation is used worldwide, particularly in urban areas, for recording the physical responses of soils and structures to natural phenomena and new civil engineering works. It enables engineers to assess the performance and impact of many sizable projects. The output is also used for recording and controlling remedial works and for assessing safety.

This book provides easily accessible guidance on a broad range of instrumentation, dealing with the conceptual philosophy behind the use of instruments, and presenting systematic coverage of their practical use. It discusses the use of displacement-dominated systems and load- or stress-based systems, and outlines their limitations and their relative advantages and disadvantages. The relevant theoretical soil mechanics background is presented in some detail and provides a simplified model against which the data can be assessed. The book also includes the basic concepts behind advanced electronic techniques, such as laser scanning in surveying and fibre-optics, with up-to-date coverage of the rapidly changing communication and data recovery systems.

It is written for senior designers, consulting engineers and major contractors who need a comprehensive introduction to the general purpose, availability and analysis of field instruments before details of their own project can be progressed. It also serves as a textbook for any specialist geotechnical MSc or professional seminar courses in which instrumentation forms a major part.

Richard Bassett is former head of Geotechnics at University College, London. He has extensive industry experience as lead consultant to the itmsoil group.

A Guide to Field Instrumentation in Geotechnics

Principles, installation and reading

Richard Bassett

Spon Press
an imprint of Taylor & Francis

First published 2012
by Spon Press
2 Park Square, Milton Park, Abingdon, Oxon OX14 4RN

Simultaneously published in the USA and Canada
by Spon Press
711 Third Avenue, New York, NY 10017

Spon Press is an imprint of the Taylor & Francis Group, an informa business

British Library Cataloguing in Publication Data
A catalogue record for this book is available from the British Library

Library of Congress Cataloging in Publication Data

Bassett, Richard, 1938–
A guide to field instrumentation in geotechnics : principles, installation, and
reading / Richard Bassett.
   p. cm.
Includes bibliographical references.
1. Engineering geology–Instruments. 2. Earthwork–Instruments. I. Title.
TA705.B325 2012
624.1'510284–dc23   2011021558

ISBN 978-0-415-67537-6 (hbk)
ISBN 978-0-203-80924-2 (ebk)

Typeset in Sabon
by Cenveo Publisher Services

MIX
Paper from
responsible sources
FSC
www.fsc.org   FSC® C004839

Printed and bound in Great Britain by the MPG Books Group

10 06233285

Contents

Figures and tables

Table

Foreword

Some years ago I was discussing with Dr Bassett the apparent lack of specific, detailed knowledge on geotechnical and structural instrumentation in the civil engineering industry. He immediately retorted, 'What do you expect! On a four-year M.Eng. course we probably teach one morning on instruments'. This made us both think about how we could address this.

Textbooks that existed tended to be either extremely over-complex, and to concentrate on a single instrument type or scheme (e.g. dams), or were little more than a collection of manufacturers' literature. No one seemed to have produced a concise guide, written by an engineer for engineers, detailing why, how and where instruments should be used.

The more we talked, the more we realised what a mammoth task producing such a book would be, so we left it firmly in the 'pending' pile. When Dr Bassett became Emeritus Reader in Geotechnics at University College London in 2006, the opportunity came to re-visit our ideas and serious work began, we sincerely hope you will find the result useful.

Chris Rasmussen
January 2011

Acknowledgements

I commenced my post-university career as a structural engineer working on an arch dam and its associated works, this included having responsibility for concrete testing in the site laboratory. Subsequently, I was instructed to develop soil testing, which included difficult plate bearing tests. A disagreement with Professor A. Bishop on the interpretation of these tests resulted in me investigating the broad soil concepts being developed by the late Professor K. Roscoe and his research group (including A.N. Schofield and C.P. Wroth), which in turn led to me undertaking a doctorate in critical state soil mechanics in 1968. My primary acknowledgement is therefore to Roscoe, Schofield and Wroth for introducing me to fundamental soil modelling and conceptual thinking; and in particular the detailed monitoring of displacements within laboratory-modelled soil structure interaction problems (these laboratory models included the full monitoring of surface stresses with load cells).

My subsequent industrial and university career involved continuous experience in investigating deformation in the laboratory and on the centrifuge. I must thank the EPSERC for x-ray facilities, Cambridge University for the use of their centrifuge, the NERCU, Balfour Beatty and Mott MacDonald for involvement in numerous real instrumentation projects. These included the Docklands Light Railway extension running to Bank Underground station, London, and the subsequent monitoring of the Mansion House project, London, and the Heathrow Express trial tunnels. I would also like to thank University College London for 'adopting' Adrian Chandler and myself, and Bristol University for enabling us to continue our displacement monitoring of structures under earthquake loading.

Following the Mansion House project I remained in close touch with itmsoil group, and as a result of the above background was persuaded by Chris Rasmussen, who had suffered through many previous arguments and debates with me, to produce this publication. Chris and Caroline Rasmussen gave me unstinting support in organising the production and proofreading of the many drafts. I also express my gratitude to itmsoil's Lucie Williams for turning my sketches into what I hope you will agree are impressive figures. Lastly, I thank my long-suffering wife for surviving my many days of absence working away at itmsoil's offices in darkest Sussex.

Introduction

This book will introduce readers to the broad concepts of field instrumentation of geotechnical construction and of the structures affected by the work. The fundamental concepts and background behind instrument types will be presented, but not as a catalogue of every manufacturer's specifications. This is because new developments are now coming forward in rapid succession, together with rapidly improved accuracy, automatic reading, and data transmission and logging. Moreover, details of specific instruments and their performance are available from all major manufacturers' websites.

The interpretation of instrument data will be discussed and, as this depends on a basic understanding of theoretical soil behaviour, Chapter 2 will outline the key relationships between soil packing, stress level, soil strength and loading paths, and pore water pressure development based on a basic form of the 'critical state model' (not on the more advanced versions such as 3SKH).

Analysts and designers are recommended to read the recent and current research available in specialist journals for a more thorough understanding of the theoretical soil behaviour models that can be used in the numerical prediction of instrument behaviour.

Chapter 1

The general philosophy of geotechnical instrumentation

Philosophy

There are several basic reasons for the instrumentation of soil constructions and soil/structure interaction problems:

a) Soil and water are the two most significant natural masses involved in civil engineering, approximately 20 kN/m^3 and 10 kN/m^3 respectively, and loads generated can be very large.

b) The behaviour of soil is the most complex of that of all materials encountered; in particular, soil properties are not under the engineers' control and their characteristics have to be assessed for every site. Soils vary in mineral content, particle size, particle interaction, packing, *in situ* stress conditions, shear strength, interaction with water, and long- and short-term behaviour. They are further complicated by possible cementing or prior failures. Every site poses individual problems, and obtaining sufficient parameters from current site investigation techniques leaves considerable uncertainty.

c) The limitation of 'all embracing' constitutive models for predicting soil behaviour means analytical methods leave rather wide areas of uncertainty.

d) Based on the currently limited site investigation data the deformations predicted for complex shear and normal loading have a wide range band, and even the most optimistic predictions are usually incompatible with the displacement sensitivity of modern structures.

e) The design engineer and his contracting engineering colleagues seek two goals in their design and subsequent construction:

   i)   As economical a solution as possible; and
   ii) Safety in both the short and long term

The interaction of the two criteria above is complex; economies involve simplicity, novelty, innovation and speed of construction with clear operating space.

Low safety factors are often associated with the above variables and the uncertainties of theoretical prediction mean economies could push a safety assessment into the uncertainty area.

The use of on-site, real-time observations of real physical changes and rates of change can give designers the ability to improve their theoretical assessments and give contractors confidence that their construction procedures remain safe.

It is to these ends that field instrumentation, of both soil projects and their associated structures, is installed.

Levels of instrumentation

To meet the various requirements, instrumentation can record and analyse two fields of data: displacements and strains, and loads and stresses.

Displacements and strains

Displacements are real tangible changes that are recorded from the datum state prior to the commencement of any work. The datum state, however, is unlikely to be an absolutely unloaded or neutral situation. Strains are, of course, differences in displacement over a period of time between two adjacent points, they are almost inevitably measured as linear strains, shear strains being difficult to identify and measure independently, although they can be assessed from three linear strains.

Loads and stresses

Stresses are in two forms, normal stress and shear stress. Stresses are a theoretical idealisation concept, i.e. a mathematical averaging in soils. They are an average loading over a specific normal area, whereas in reality they are the sum of a large number of forces at point contacts. Stresses in water are an acceptable concept as water is a continuum that does not carry shear.

Loads in structural members can be measured directly, or inferred from strains in material with well-defined properties such as steel. Again, the absolute zero may not be available unless the instrument is calibrated independently.

Rates of change

Both displacements and loads change with time and hence rates of change can be assessed, but the sensitivity and stability of the measuring instruments must be of very high quality and of sufficient resolution.

Types of monitoring

To meet the economic and safety criteria outlined earlier, instrumentation can fulfil a number of roles and respond at appropriate levels. The author suggests these roles are divided into three generic categories:

a) Passive monitoring, for record purposes and, ideally, later back-analysis.
b) Near real-time monitoring, for construction control.
c) Monitoring for safety, when some action or construction could potentially result in an unacceptable result or a dangerous situation developing.

Passive monitoring

Passive monitoring involves recording and storing the developing displacement and stress data to validate that the predicted safety thresholds are not being exceeded. It is a passive function where the final data patterns produced by theoretical prediction, or from previous experience are most unlikely to exceed any damage thresholds. The instrumentation data is essentially a record that provides the owner, the engineer, the contractor and interested third parties with proof that the real behaviours of the soil and the structures comply with the expectations. Within the predicted limits the situation can be considered satisfactory.

In such cases the measuring system would be used for some time prior to work commencing in order to determine the normal daily movements, and data would then be inspected and assessed once a day during the civil engineering work. In the unlikely event that deformations and/or stresses are apparently going to exceed the predicted values, there may be time to initiate remedial action on the structure or to alter the construction technique.

If this passive technique is used, the data should be published, as it provides a record of the experience.

Back-analysis of the data can significantly improve later theoretical prediction.

Near real-time monitoring

In this situation instrumentation is used when: a) it is necessary to record and positively control remedial work, such as compensation grouting or jacking; b) when the predicted stress changes or displacements are forecast to slightly exceed acceptable values. Preferably, the hardware and the recording systems would be the same as in passive monitoring, but the rate of reading the data would be increased dramatically and displayed in real time on a PC screen so that it can provide real control. Decisions can be made on how and when to intervene, and (if remedial work is in progress) when to start and stop the processes. The varying responses observed and all the data must be recorded for later reappraisal.

This approach is developed into the 'observational method' (Peck, Powderham), when the construction technique is progressively changed from a previously used acceptable technique towards a simpler, more economic and faster construction method. Monitoring becomes progressively more significant as the margins of safety are lowered. Real-time assessment is mandatory for any 'observational method' but must also be accompanied by having remedial plans and equipment in place, which can rapidly prevent a dangerous situation developing.

Monitoring for safety

The major deformation and incipient failure of such works as deep braced excavations and tunnelling are governed by the displacements and rates of closure that take place in props and behind walls, and by pore water changes near the works, during and following the sequence of construction. For safety monitoring it is not primarily the absolute data values that are important, it is the rate of change of data, though in terms of pore pressures an absolute value may be critical. To measure rate of change confidently and in a way that enables real-time decisions to be made, data have to be recorded at rapid intervals to a high level of accuracy, and the data processed immediately to provide the rate of closure in millimetres of closure per minute or rate of stress change. Displayed as a plot against time, any significant acceleration observed can then be acted upon. Such a requirement is almost identical to the real-time monitoring described above, but processing will be more local and more detailed and an experienced engineer must make crucial decisions without hesitation – this can result in a 'catch 22' situation.

To cope with this changing set of requirements the instrumentation system must be capable of automatic reading, the data from all instruments must be captured at consistent, discrete points in time, which may vary in interval from once an hour to every 20 seconds. The resulting vast quantity of data must be directly accessed and recorded by a PC and processed rapidly to provide a clear visual picture. It is impossible to assess every variable in emergency situations and early decisions must be made to identify the very sensitive, critical changes.

Previous experience must be available as a database of satisfactory and unsatisfactory performance, either from a planned trial, as at Heathrow (see Chapter 7), or from a learning curve on similar, earlier sections of the actual job. Many electronic instruments are suitable for safety monitoring, and include vibrating wire piezometers, strain gauges, linear variable differential transformers (LVDTs), servo accelerometers and electrolevels. Many instruments can be connected to their monitoring datalogger by radio (see Chapter 9), which avoids complex wiring systems, considerably shortening the installation time. They have different costs, sensitivities, stabilities and reliabilities.

Chapter 2

Basic soil mechanics and pore pressures

Prior to describing the operation of field instruments and their interpretation the author will digress to outline some of the basic behaviour of soils so that instrumentation measurements have a relevant background against which they can be interpreted and assessed.

Soils as a simplified model

The void ratio e (i.e. packing) of clays can vary enormously in normal conditions (from $e = 3 - 5$ down to $e = 0.3 - 0.5$), where $e =$ volume of void/volume of solid, see Figure 2.1.

$(1 + e)$ is called the specific volume. It is the volume occupied by one cubic metre of solid (weight approx. $2.75 \times \gamma_w = 27.5 \, \text{kN/m}^3$ of soil); if the void e is assumed to be full of water the specific volume contains $e\gamma_w = 10e \, \text{kN}$ of water.

The saturated density can range from:

Very soft clay $(e = 3)$ $\frac{27.5 + 3 \times 10}{4}$ approx. $= 14.4 \, \text{kN/m}^3$ (e.g. estuarine mud)

Dense clay $(e = 0.5)$ $\frac{27.5 + 0.5 \times 10}{1.5}$ approx. $= 21.7 \, \text{kN/m}^3$ (e.g. very over-consolidated clay)

This change is caused by increasing normal stress and results in 'consolidation', which in practical terms causes settlement. Because soil is a 'semi-infinite half space', the consolidation volume change is dominated by only *vertical movement* (settlement), see Figure 2.2.

In general, stresses are three dimensional. In civil engineering, however, most projects are *two dimensional* involving changes in stresses and displacement in the vertical and in one horizontal direction; the other horizontal direction experiencing stress changes but no displacements or strains, in effect one horizontal direction remains a K_0 direction. This situation is governed by two principal stresses σ'_1 and σ'_3, and this will be assumed throughout this simplified approach. The interrelationship of mean normal stress $s' = \frac{\sigma'_1 + \sigma'_3}{2}$ and of maximum shear stress $= \frac{\sigma'_1 - \sigma'_3}{2}$. Normal and shear stress in two dimensions can easily be represented by a 'Mohr circle of stress', see Figure 2.7 (later). The fundamental parameters of a circle are the centre and the radius, and in terms of principal stresses the centre s' and the radius t.

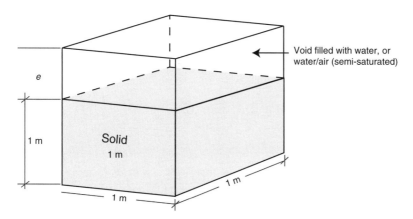

Figure 2.1 Diagram of the concept of void ratio.

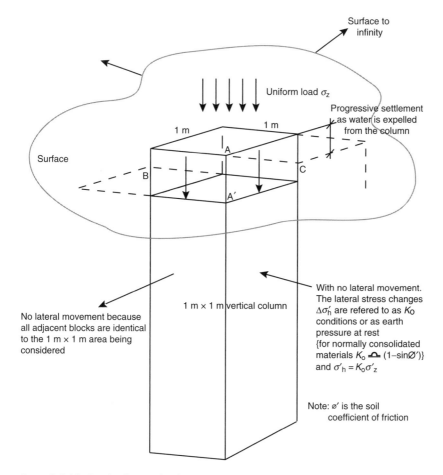

Figure 2.2 Idealised column of soil experiencing one-dimensional consolidation.

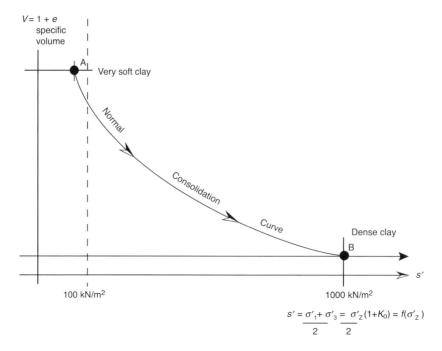

$$s' = \frac{\sigma'_1 + \sigma'_3}{2} = \frac{\sigma'_z(1+K_0)}{2} = f(\sigma'_z)$$

Figure 2.3 Normal consolidation for a clay soil on a $1 + e$/normal stress plot.

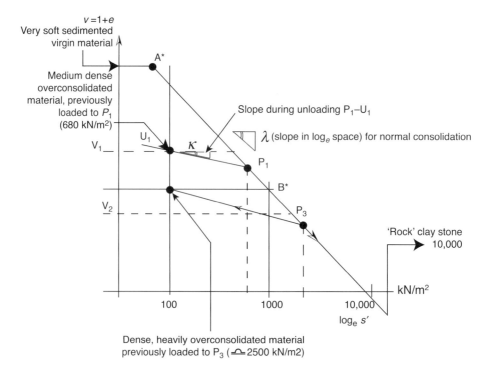

Figure 2.4 Normal consolidation for a clay soil on a $1 + e / \log_e$ normal stress plot.

Logic and experimental evidence confirms that the relationship between s' and the packing $1 + e$ is as shown in Figure 2.3.

For a K_0 normally consolidated situation

$$s' = \frac{\sigma_1' + \sigma_3'}{2} = \frac{\sigma_z'}{2}(1 + K_0) = f(\sigma_z')$$

So s' can be replaced by a simple function of vertical stress σ_z'.

This is in fact a linear/log relationship, shown in Figure 2.4.

This normal consolidation relationship only occurs on the initial 'geological' loading, If erosion unloads the material, the recovery is elastic and the slope $\mathcal{K}$ is totally different ($\mathcal{K}$ is approximately 1/10 of λ, i.e. P_1 to U_1 in Figure 2.4). This line, as it is elastic, can be reloaded (approximately) from U_1 to P_1.

Let us look at a band of material from V_1 to V_2 but on the simple v/s' plot, concentrating on one value v of packing, see Figure 2.5.

All have the same void ratio $(1 + e) = V$, yet each specimen X, 1, 2 and 3 has reached this from different original P values. The other variable in clay is its shear strength and its response to different forms of loading, depending on a number of variables. A general two-dimensional loading case is shown in Figure 2.6. The small two-dimensional rectangular element a–b–c–d is subjected to normal and shear stresses on each surface; they are in equilibrium. Some basic trigonometry assuming equilibrium on any plane across the rectangle will show that there are two planes (the principal planes) on which there are no shear stresses and that the stress pairs on any plane $\sigma_\theta' : \tau_\theta$ can be plotted as a circle, see Figure 2.7. The principal direction shown in Figure 2.7 is

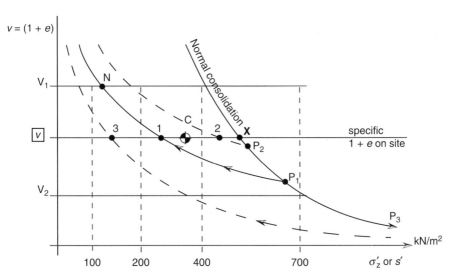

Point X would be called *A normally consolidated clay*
Point 2 would be called *A very lightly over-consolidated clay*
Point 1 would be called *An over-consolidated clay*
Point 3 would be called *A heavily consolidated clay*

Figure 2.5 Unloading (swelling) curves on a $1 + e$/normal stress plot.

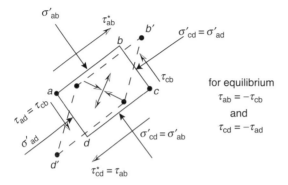

Figure 2.6 Equilibrium stresses on a rectangular element.

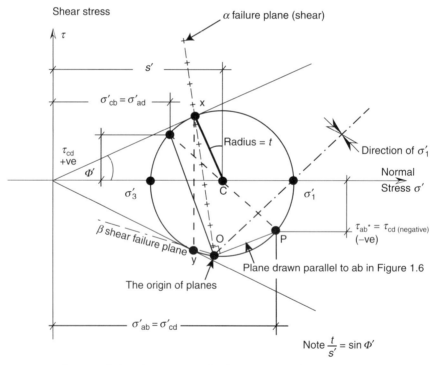

Figure 2.7 'Mohr circle of stress' at failure.

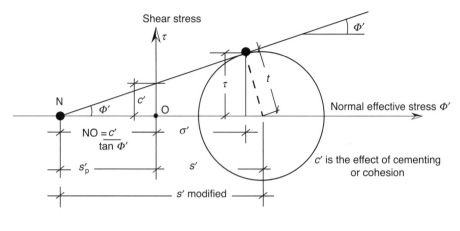

Modified mean normal stress $= \dfrac{\sigma'_1 + \sigma'_3 + s'_p}{2}$

Maximum shear stress remains $= \dfrac{\sigma'_1 - \sigma'_3}{2}$

Figure 2.8 'Mohr circle of stress' for a drained material with cementing or cohesion.

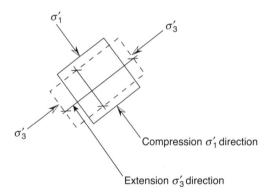

Figure 2.9 Deformation in the principal stress directions.

a key piece of information for the understanding of load distribution in soils. It can be found by identifying the 'origin of planes' (O in Figure 2.7). This is found by identifying one stress pair $\sigma'_{ab} : \tau_{ab}$ then drawing the line PO parallel to the plane a–b in Figure 2.6.

Once the point O has been identified, two other key planes can be defined: O_x and O_y, which represent the 'planes of maximum obliquity' or the shear failure planes α and β within the soil element on which

$$\frac{\tau}{\sigma'} = \tan \Phi'$$

Many real soils have a cementing or cohesive component which is often shown by intercept c' in Figure 2.8. In analytical terms this component is more easily treated as a pre-stress s'_p added to the mean normal stress s', as the intercept $s'_p = \frac{c'}{\tan \Phi'}$, and analysing it as a simple frictional material based on a modified mean normal stress $s'_{\text{modified}} = s' + s'_p$.

This maximum shear stress is critically important because, as water is a fluid, *it can take no shear*, so

$$\frac{\sigma'_1 - \sigma'_3}{2} = \frac{\sigma_{1 \text{ total}} - \sigma_{3 \text{ total}}}{2}$$

the shear stress t is carried *completely by the soil structure*.

Therefore, t_{max} is a key property of the soil. Using the principal directions, the deformations are simplified as there is no shear. Although both principal stresses are compressive, the strains in the σ'_1 direction are compressive but those in the σ'_3 direction are tensile, as shown in Figure 2.9.

As soil particles are more-or-less independent units their points of contact are dominated by: i) friction related to normal load, and ii) small surface electrochemical or cementing at contacts.

If the shear stress at contact points exceeds the cementing and frictional capacity, the dominant effect will be slip (irrecoverable), i.e.

(i) $T_{(i)} = N \tan \Phi' (\Phi' = $ angle of friction); and
(ii) $T_{(ii)} = c'$

c' can be written as $N_p = \frac{c'}{\tan \Phi'}$ i.e. an equivalent pre-stress N_p (see earlier) giving $T = (N_p + N) \tan \Phi'$ or $t = (s'_p + s') \sin \Phi'$ (don't worry, study Figure 2.8).

If *one* elastic unloading line is treated as a unique structural packing undergoing elastic unloading or reloading $(P_1 - 1)$, see Figure 2.5, it is suggested that there is a limiting 'state boundary' relationship between the maximum shear stress t and s', inside which the soil is elastic, outside which the soil structure cannot exist but on this state boundary the soil plastically yields. This boundary can be represented on a t/s' plot above the swelling line P_1–1–N (Figure 2.5) as in Figure 2.11. This concept is essentially the key to the state boundary surface in the 'critical state model' for soils.

Figure 2.11 represents the yield envelope associated with different forms of yielding. There are, however, an infinite number of these boundaries, one above every swelling line, and they define the $t - s' - 1 + e$ surface shown in Figure 2.14 (later). There are many simple mathematical forms of this boundary, the simplest probably being the ellipse of the Roscoe–Burland model. Figure 2.12 shows this elliptic form and the assumed normality vectors of plastic strain.

The plastic strains have two components, volumetric strains (ε_v) associated with changes in s' and shear strain (ε_γ) associated with changes in t. The key point is (ε_v) at point ©, where under the normality rule there is pure shear strain with no volumetric strain and no changes in t or s', i.e. the material flows like a liquid, this is the 'critical state point' for this particular elastic swelling line.

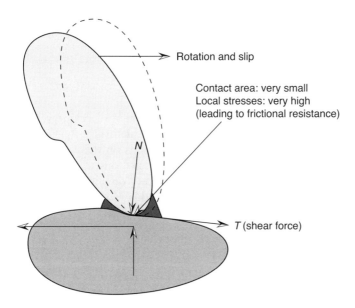

Figure 2.10 Idealised grain contact.

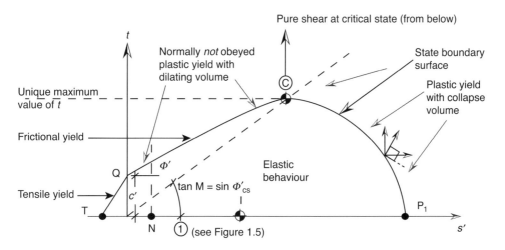

Figure 2.11 State boundary relationships for a clay on an elastic unloading line.

C is the maximum t value, which for the assumed ellipse occurs at $\frac{s'_p}{2}$, the critical state condition. When shear loading reaches this point the material fails with *no* volume change, i.e. as above it behaves like a viscous fluid (flows). The surface $p'–c–q$ is a state boundary surface *and* represents a yield surface, i.e. when a soil is loaded (t and s' changing) if the value is inside the boundary (and its reflection is on the s' axis) it will behave elastically. If the stress reaches the surface it will behave plastically, and thereafter stress changes are confined to the boundary surface.

It must be remembered that there are an infinite number of yield surfaces, one above each individual unloading line (e.g. Figure 2.5, P_3–P_1–P_2, giving Figures 2.13 and 2.14).

The key points are C_2–C_1–C_3, which in the t/s' plot *all lie on a straight line – the critical state line*. i.e. C_2–C_1–C_3 ($t/s' = \sin \Phi'_{\text{critical state}}$).

With an infinite number of yield surfaces a complex three-dimensional surface in the $t - s' - 1 + e$ plot is produced, see Figure 2.14.

The hatched vertical surface above P_1–①–N forms a vertical wall (Figures 2.11 and 2.12) and is termed the elastic wall. On the curved, vertical surface P_1–C_1–q_1–N–①–P_1 the material behaves elastically (like steel) and remains on this wall if any changes of t and s' occur together with complete drainage, i.e. no constraint on $1 + e$.

Why all this complexity? The interpretation of piezometers on instrumentation projects can be better understood within the framework of this model, particularly for undrained loading of clays and the subsequent changes with time.

Pore pressure measuring instruments

The key use of the critical state concepts within field instrumentation is to assess the consolidation and strength changes that occur during a project. The fundamental key to these changes in the normally saturated soils of the United Kingdom is the careful monitoring of the development and decay of pore water pressures. A simplified guide will be presented, which it is hoped will clarify some of the quite complex responses that are observed. First, however, the basic units used for this purpose will be described.

Water is a continuum, a fluid, and theoretically is incompressible; it also exerts a uniform pressure in all directions. A unit capable of being inserted in the ground at a point will record this hydraulic pressure and any changes on a pressure-sensitive surface, provided it requires very little volume change to activate the surface and is not in contact with the soil particles. The measuring unit is therefore installed in its own small water reservoir protected from the soil by a stiff porous barrier, see Figure 2.15.

Unfortunately, real water is not a completely degassed fluid and inevitably contains dissolved gases; changes in water pressure, particularly reduction or even suction, will cause gases to come out of the fluid in the form of fine bubbles. Changes in pressure with gas bubbles present cause considerable volume change and, if these are within the unit's own reservoir, it may take a considerable time to come to a new equilibrium.

The porous tip in such circumstances must be fine enough to eliminate or minimise air entry from the soil. If the ingress of air cannot be eliminated satisfactorily then the small reservoir chamber may need to be flushed with de-aired water using a flow and return pipe, see Figure 2.16 – a geo-flushable piezometer.

Types of piezometer

There are a number of measuring units available which are capable of quantifying the uniform water pressure u on the measuring face, all require some small displacement of the face, these are described below.

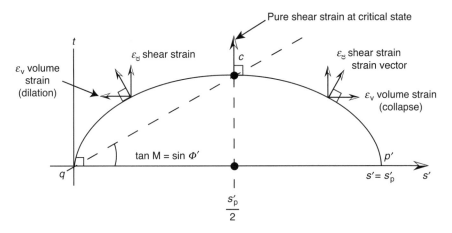

Figure 2.12 The simplified ellipse form assumed in the Roscoe–Burland version in the CRISP program.

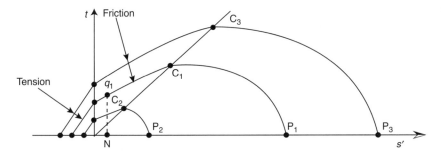

Figure 2.13 Set of state boundary surfaces.

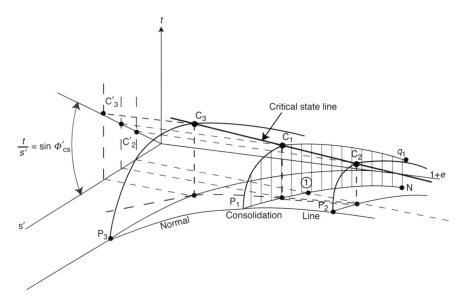

Figure 2.14 Isometric view of the state boundary surface on s', t, $1 + e$ axes.

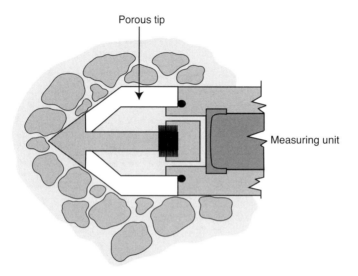

Figure 2.15 Simplified pore pressure measuring unit.

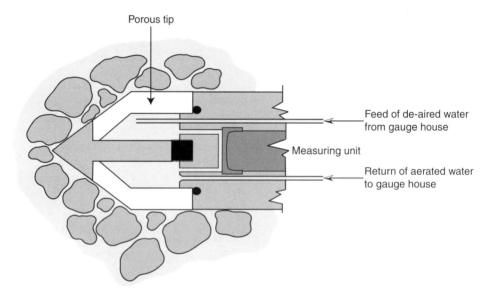

Figure 2.16 Flushable modification to a piezometer.

The pneumatic piezometer

The early most-direct system was to cover the measuring face with a thin flexible diaphragm, pressed against a flat metal block by the pore pressure. Two holes in the block connect to pipes to the readout unit. Gas pressure (normally nitrogen gas) was passed down one pipe, this pressure being slowly raised until the applied gas pressure lifted the flexible diaphragm and the gas returned via the second pipe to the readout

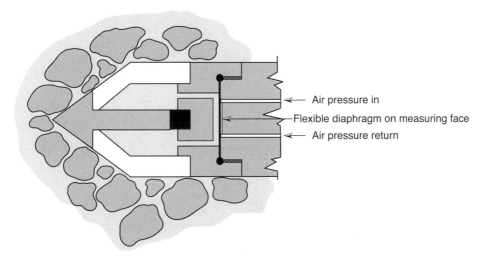

Figure 2.17 Schematic diagram of a pneumatic piezometer.

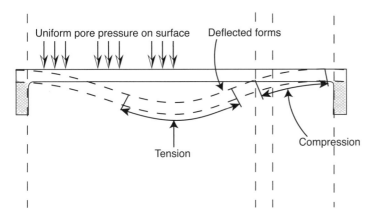

Figure 2.18 Deformation of a diaphragm.

unit. This 'crack open' pressure was recorded noting the reading of a pressure gauge when the return gas flow appears.

The applied gas pressure was then slowly lowered until the pore water pressure in the ground u assumed to be in the piezoemeter unit's reservoir re-shut the diaphragm and the return gas flow ceased, again the pressure gauge reading was noted, and the procedure repeated several times to determine u (Figure 2.17).

Strain-gauged diaphragm piezometer

This measuring unit consists of a thin metal (stainless steel) disc with fixed edges. In this case, the uniform pore pressure u deforms the diaphragm as shown in Figure 2.18.

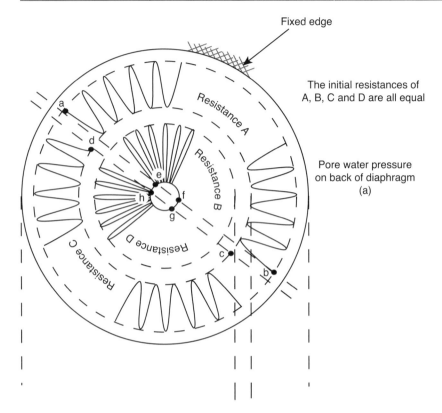

Fixed edge

The initial resistances of
A, B, C and D are all equal

Pore water pressure
on back of diaphragm
(a)

Figure 2.19 Arrangement of a diaphragm.

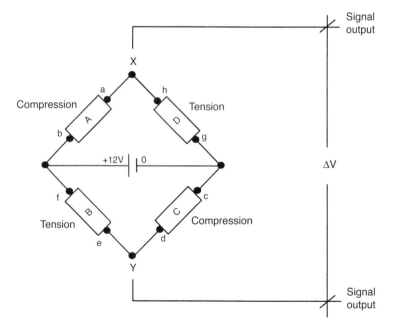

Figure 2.20 Simplified full Wheatstone bridge circuit.

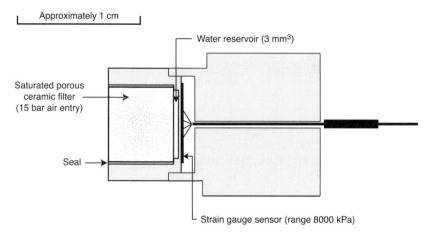

Figure 2.21 Suction gauge (after Imperial College, London).

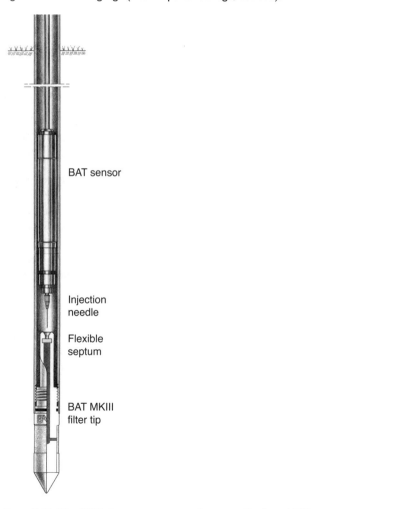

Figure 2.22 The BAT piezometer system (courtesy Profound BV).

Currently the commonest use of the diaphragm is with a vibrating-wire system, which is described in detail in Chapter 4.

An alternative to the vibrating wire is the use of a special etched strain-gauge unit attached to the back of the diaphragm. As shown in Figure 2.18, the diaphragm experiences a double curvature: the outer band experiencing concave bending (compressive surface) and the centre experiencing convex bending (tension surface), the etched gauge is shown in Figure 2.19.

The four resistance blocks shown in Figure 2.20 have the same resistance and are arranged as a full Wheatstone bridge circuit. This arrangement produces increased resistance on the tension surface and lower resistance on the compression surface. The result being the voltage at X increases above 6 volts (assuming a 12-volt supply as in the example illustrated) while at Y it falls below 6 volts. This voltage difference can be measured on a millivoltmeter and, as the circuit is a full bridge (i.e. it is balanced), it is automatically compensated for temperature variations.

This principle is also used in a miniature suction pore pressure probe developed at Imperial College, London, shown in Figure 2.21. The key features being the high 15 bar air entry ceramic tip and the minimum reservoir volume of only approximately 3 mm^3.

Removable piezometers

Piezometers are often required to operate reliably for very long periods of time (for example, more than 40 years for dams) and there are often questions and concerns over drift or a breakdown of the measuring unit. Many modern piezometers have proven stable and repeatable over these long periods, but if there are any serious doubts then a removable measuring unit can be considered.

An example of such an instrument can be found at Profound BV in The Netherlands, who produce a BAT piezometer system that consists of an externally applied measuring unit fitted with a hypodermic needle, see Figure 2.22. The piezometer tip is installed at the end of a 25 mm diameter pipe connected to the ground surface. The tip chamber is sealed by an elasto-polymer diaphragm, which can be penetrated by the hypodermic needle many times and yet automatically reseal. The key advantage of this system is that the measuring unit can be removed and recalibrated or a new unit installed.

Using the simplified model to interpret pore water pressure responses

When a clay is loaded, the civil engineering work typically takes two months to two years. Even two years can be too short for the clay to drain (its permeability being 10^{-7} to 10^{-10} m/s). Water is ideally an incompressible fluid so its *volume cannot change*. If water cannot get out (drain) then the volume of the whole saturated soil mass *cannot change*, i.e. $1 + e$ *remains constant*. In the sections above on generalised soil behaviour the relationship between $1 + e$ (the specific volume), s' (the mean normal stress), and t (the maximum shear stress) has been outlined, giving rise to the concept of a state boundary surface, the critical state line, the zone within which behaviour is elastic, and when and what components of plastic yield occur on the state boundary surface. Civil engineering work subjects soil elements to changing, often complex,

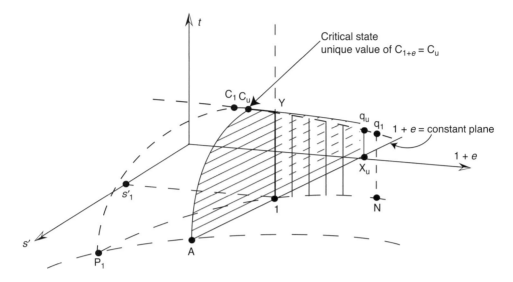

Figure 2.23 Undrained surface ($1 + e =$ constant).

stress conditions. In the $t - s' - (1 + e)$ diagram, the undrained situation is represented by one unique plane ($1 + e$ constant), this is shown as the plane A–C_u–q_u–X_u–1–A in Figure 2.23.

If, for example, an element of soil starts its *in situ* stress state at 1 lying on a specific elastic wall P_1–1–N and is subjected to a shear loading with t increasing but s' constant (the complex concept will follow later), as the soil is elastic, it must climb up the elastic wall but it must also remain at $1 + e =$ constant, which in this specific circumstance it does until point Y as this is the only common path for both $1 + e =$ constant and the elastic wall. At Y the path cannot go on upward because the soil cannot exist outside the state boundary surface, therefore if t goes on increasing, the effective soil stresses are forced to move from Y to C_u. At C_u catastrophic flow failure will occur still at a packing of $1 + e$.

The path 1–Y–C_u is the effective stress path for all undrained loading to failure from 1. This was a specific case, let us look at some more general or typical cases.

What total loadings do civil engineering works produce? These loadings are termed 'applied stress paths', best illustrated by elements in a number of typical situations.

A cantilever retaining wall to be built at (a–b–b′–a′) then excavated on the right-hand side to c–d in Figure 2.24: the typical locations of a number of piezometers are shown at element points A, B, C, D and E. The void ratio is assumed constant throughout, i.e. all the elements therefore lie on *one* undrained plane but their individual s'_1 values (i.e. their start location 1) are different.

Each element on the $1 + e$ plane is shown in Figure 2.25 (the spread of points is grossly exaggerated) on the simple assumption that initially the conditions are isotropic, i.e. $\sigma'_z = \sigma_h$, no shear stresses exist (this is not very realistic as over-consolidated clays have horizontal stress between $0.8\sigma'_v$ and $2.5\sigma'_v$). In some cases advanced site investigation has provided the real values of K_0 where $\sigma'_h = K_0\sigma'_2$ and using a Camkometer (also

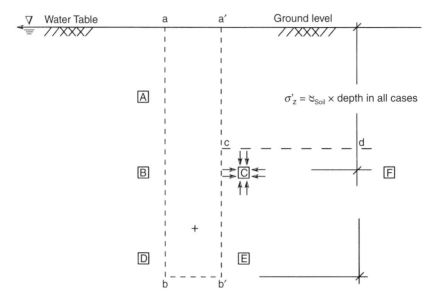

Figure 2.24 Detailed wall section.

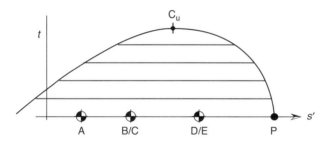

Figure 2.25 Locations of initial total stress for A to E.

known as a self-boring pressure meter) instrument and the real s' and t values *in situ* can be used in the assessment.

If after the wall a–a'–b'–b is installed (and assumed to be installed so that no stress changes occur) the material from ground level to c–d (foundation level) is then excavated. How does the wall move? Figure 2.26 shows an exaggerated displacement diagram (inclinometer data).

So what happened to the total stresses on the elements?

1a) The total vertical stresses (σ_z) on A, B and D all remain the same.
1b) The total horizontal stresses (σ_h) on A and B fall as the wall moves away *but* the horizontal stress on D rises due to the point of rotation of the wall.
2a) The vertical stresses on both C and E *fall* by the weight of soil removed by excavation (i.e. h to c–d).

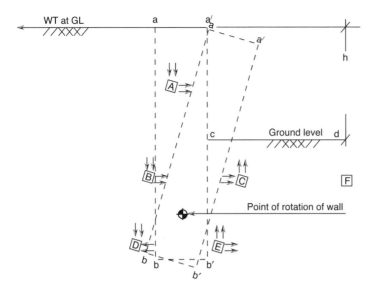

Figure 2.26 Magnified rotation of the wall due to excavation.

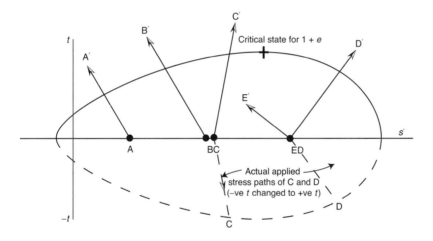

Figure 2.27 Total stress paths A to E.

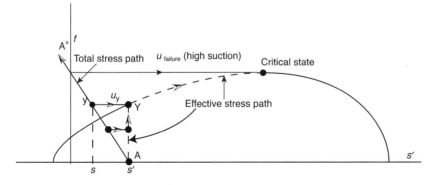

Figure 2.28 Stress paths for element A.

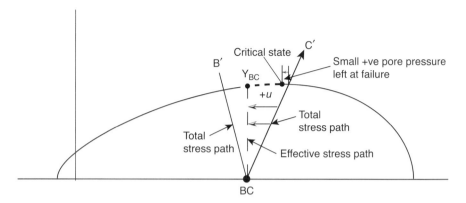

Figure 2.29 Stress paths for elements B and C.

2b) The horizontal stress on C climbs as the wall pushes on element C but the horizontal stress on E also falls due to the point of rotation.

2c) For element C, σ_z falls noticeably and σ_h rises massively, therefore s_{total} rises while t rises rapidly. The σ_h falls from C to the point of rotation (PoR) depending on depth, as noted t rises massively at level C but the rise becomes less dramatic with increasing depth from C to the PoR.

2d) For E, σ_z falls as for C above but σ_h also falls, therefore s_{total} falls quite sharply while t increases slightly.

These are changes in applied stresses. What do these various stress paths look like on the t–s plot?

Figure 2.27 shows the various *total stress paths* that each element will experience. If for convenience we reflect the paths C–C, and D–D onto the positive t side of the s axis, i.e. C–C' and D–D', the actual change in t is negative.

These are 'applied total stress paths'. Now *all* stress paths experienced by the soil structure itself are in terms of *effective* (soil) *stresses*. These *must* follow the constraints shown in Figure 2.23 and all must travel vertically upward with s' constant until the yield surface is reached at the appropriate y value (see Figure 2.23, the s' value remaining unchanged), thereafter the effective stress path must travel from Y to $C_{critical\ state\ failure}$.

How does the total stress path relate to the soil's effective stress path?

Changes in pore water pressure cannot affect the shear, so it modifies s_{total} to s', see Figure 2.28. It is assumed for simplicity that the total stress path and the effective stress path both start from point A. In effect the total stress path is offset by the static ground water pressure.

The total path is shown, A–A*. To modify the total stress path from A to Y, the effective stress from A to Y can only be achieved if s has been increased to s', so a *suction* (or negative pore water pressure) is generated in the soil u_y. From Y to the critical state a huge increase in negative pore water pressure is generated reaching $u_{failure}$. In practice, the soil even gaps away from the top of the wall and cracks.

Now consider elements B and C (see Figure 2.29).

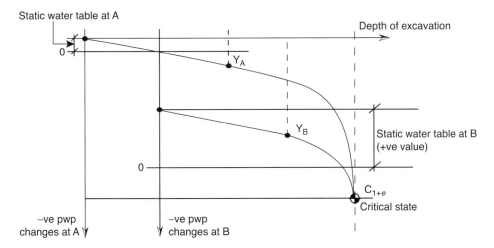

Figure 2.30 Changes in pore water pressure (pwp)/depth of excavation.

The total stress path B–B' has the same form as A above, in that the 'effective stress' path has a higher value of s'. Again the effective stress Y_{BC} is higher than the total stress path so pore pressure suction will be generated, and after Y it will increase faster until critical state is reached. The changes in pore pressure for elements A and B are shown relative to the excavation depth in Figure 2.30, but in this case the static water level has been included. The absolute pore water pressure values measured on field instruments must indicate the static water head.

In contrast, the total stress path C–C' in Figure 2.29 is higher than C–Y_{BC}, hence the element will initially show the generation of the positive pore water pressure in a very localised area in front of the wall as the yield condition approaches; but from Y_{BC} to critical state, as ultimate failure develops, the positive pore pressure may well start reducing to zero and in some cases may become negative at failure.

Element E (see Figure 2.31) will undergo a total stress path with s falling rapidly and t rising slowly (E–E'), the effective stress path will still be E to $Y_{d/e}$ and on to the critical state. Hence a very rapid generation of negative pore water pressure will occur in a localised area round the toe; however, t is unlikely to rise out of the elastic behaviour of the range. D on the other hand will generate high positive pore water pressure behind the toe of the wall, with an increasing rate as failure approaches from Y_d to critical state as the total stress rises in the last stages to D', but the effective stress path is still D–$Y_{d/e}$–C. It should be noted that very marked suctions and pressures occur around and above the toe of a cantilever wall. Local redistribution of water pressures and softening may occur more rapidly than anticipated.

Well away from the wall at F (Figure 2.24) the removal of vertical stress will predominate. The horizontal stress will fall as a K_0 unloading situation, t will thus slightly increase in a negative sense while s falls, high negative pore water pressure will develop. As time and dissipation of the suction occurs, softening and heaving of the clay in the base of the excavation will result.

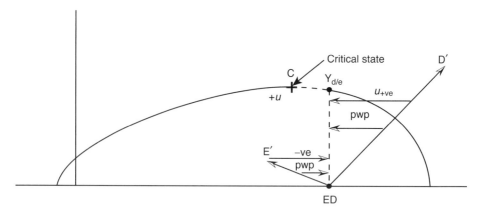

Figure 2.31 Stress paths for elements D and E.

Cuttings

Figure 2.32 shows the formation of a cutting in clay. Initially all total stress paths have falling s and rising t, this will generate suction pore water pressure to maintain the original s' effective pressure as seen in Figure 2.33. Water will be sucked into the section mainly from the ground water table and from the surface, resulting in swelling and weakening of the slope surface and of the formation. Long-term surface slipping is common in railway and motorway cuttings. If the slope b–c is not cut at the critical – state Φ' value a long-term creep failure is almost inevitable. This can be reduced even further if steady-state seepage develops down the slope. Surface drainage of clay cuttings is essential.

In Figure 2.32 possible locations of piezometers are marked A_1, A_2, C_2, C_4, etc. A failure mechanism (slip circle) may develop within the zone a–b–v–c–d–w–n–a. The greatest shear strain will be in the upper section a–b–v–n where cracks and settlement steps may occur (safety of the slope being located in the c–d–w zone).

Do not dig a ditch in the c–d area without carrying out a safety analysis and backfilling with strong granular material.

Note that nearly *all* the applied stress paths for the piezometer locations show falling values for s_{total}. The increasing horizontal stress in the area E_6–E_8–D_9–D_5 is really the most interesting as it may compensate for the vertical total stress falls giving rise to rapid rise in t with little change in s_{total}, see path D_7. If there is no change in s_{total} then no change in pore water pressure will develop prior to yield; changes will develop only after yield as the critical-state failure is approached, giving little warning. Pore water suction will be generated throughout the remaining section, the key zone (A_2 to D_5). In the lower part of the shear zone (piezometer D_5 to D_7), the response will be as in Figure 2.33. If E_6 is at depth it may go into positive pore pressure at failure, again see Figure 2.33.

This brings us to a key consideration for this genre of problems; what happens as water is slowly sucked into the soil (from rain on a–b–c–d) and from seepage through the general soil? Piezometers A_1, A_2, C_4, D_7 and E_6, etc., will all register a progressive modification to a new steady-state seepage condition with water pressures lower than

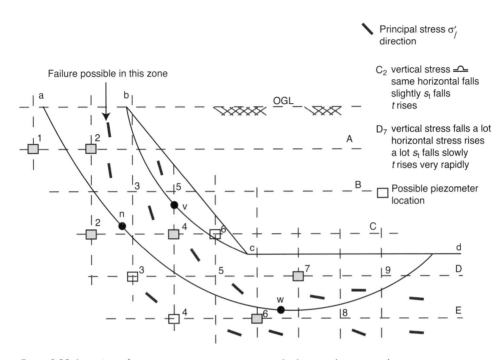

Figure 2.32 Location of pore pressure measurement units in a cutting excavation.

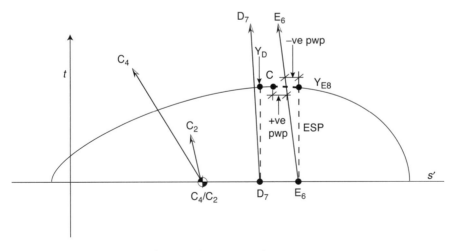

Figure 2.33 Total stress paths for typical piezometer locations.

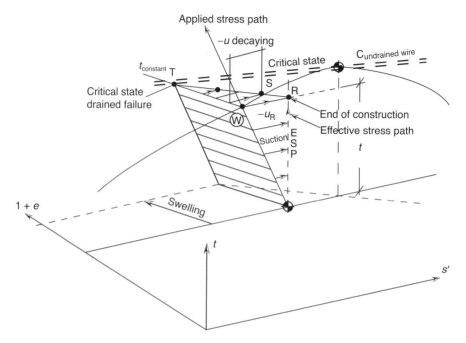

Figure 2.34 Decay of pore suction with swelling of the clay and fall of the critical-state strength.

the original static water levels. The soil will swell and the void ratio will therefore increase; the value of the relevant critical state point C (maximum shear strength) will progressively fall. Failure can eventually follow months or even years after completion of the project; a common experience in motorway cuttings. The owner needs to maintain long-term records.

Figure 2.34 shows the behaviour diagram in three dimensions $(1+e, t, s')$, the total stress path assuming an undrained safe condition initially rising to a value of t at point W, the effective stress reaching R, with $-u_R$ as the suction pore water pressure. The value of t does not change so the effective stress moves R to S to T, a critical state where drained failure occurs as $1+e$ swells.

Embankments

Embankments are a two-fold problem, the foundations are a natural soil (which is probably inhomogeneous and anisotropic). Whereas the embankment, even if made of the same material (clay), will have been excavated, moved, reconstituted and rolled. It will probably be less dense but will now be a semi-saturated, isotropic remoulded material and will behave somewhat differently to the foundation material.

Figure 2.35 shows potential locations for piezometers in the foundations; the applied stress paths of D_1, D_5 and D_9 (inverted in terms of a negative t value and a negative C) are shown in Figure 2.36. Points of interest are that very high positive pore water pressures developed particularly at D_1. At D_5 the total stress path generates positive

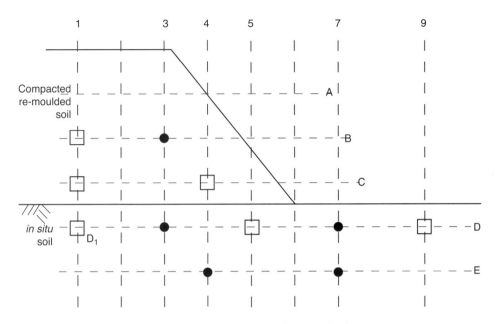

Figure 2.35 Locations of pore pressure measuring units for an embankment construction.

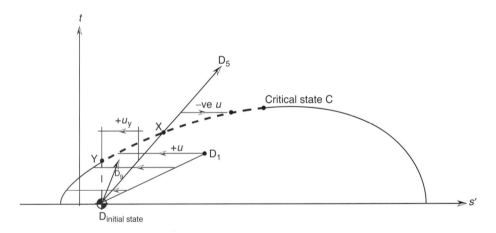

Figure 2.36 Total effective stress paths for the D_1, D_5 and D_9 units.

values at first from $D_{initial}$ to Y and subsequently from Y to X, they decay to zero at X then gradually increasingly become negative until failure occurs at C.

A similar result will occur with D_9; t is not likely to be raised above the equivalent zero pore pressure point. The critical area is piezometers D_3 to D_6 where, if the pore water pressures remain positive during construction, they will subsequently decay with the material consolidating to a denser and stronger packing. Within the embankment material the compaction will remould the material, lowering the $1 + e$ value. Additional compaction and the weight of more layers of applied soil can result

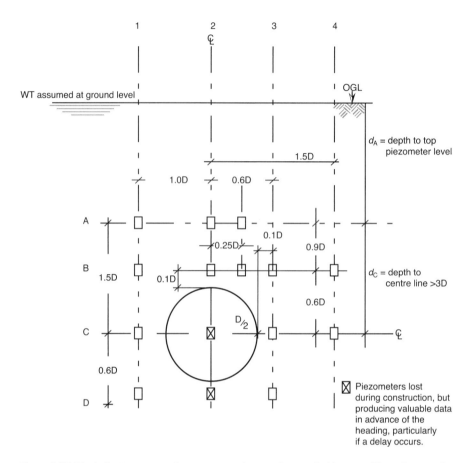

Figure 2.37 Typical piezometer layout around a trial tunnel (4 m to 10 m diameter) in over-consolidated clay.

in high pore water pressure generation, reaching, in extreme circumstances, the full dead weight of the soil added; the critical layer will then act like a water bed. This generates no extra shear strength and the soil becomes a confined viscous fluid. Embankments suffering this problem bounce and roll with a wave motion when crossed by machinery.

Tunnels

Construction of new tunnels in saturated over-consolidated clay materials relies during construction on the fact that the opening can be formed without support for a short length of time. This is termed by miners 'the stand up time'; it can vary from minutes to many hours. The key is the development of major falls in pore water pressure, even resulting in suction pressures, these are maintained for some time due to the low permeability of the clay, which in ideal conditions (i.e. a homogeneous material with no permeable cracks) decay slowly. The slow decay can be ruined by the presence of sand

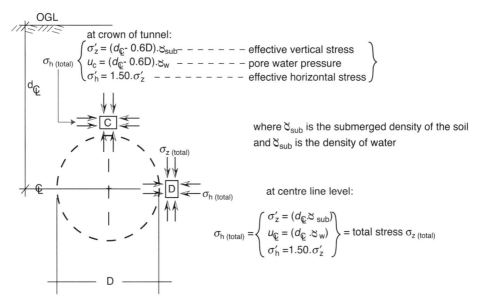

Figure 2.38 Initial stress conditions prior to tunnel construction in an over-consolidated clay.

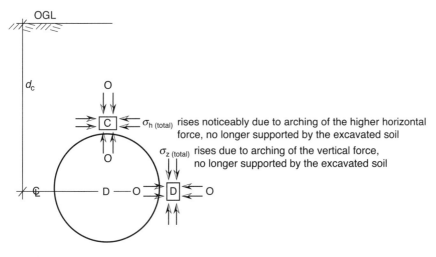

Figure 2.39 Total stress conditions after tunnelling.

partings and fissures in the clay, allowing much more rapid decay. The careful measurement of pore water pressures and their decay rate is therefore critical to planning for both the construction sequence and safety within the tunnel heading.

Figure 2.37 shows the ideal layout used for a trial tunnel section. Monitoring in the final permanent construction would be simpler, as it can be based on the detailed data acquired in the trial.

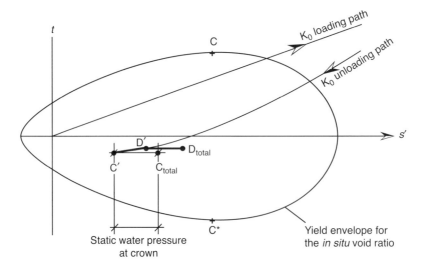

Figure 2.40 The development of over-consolidated stresses in geological time to give the stress conditions at points (C′ and C) and (D′ and D) in the original ground.

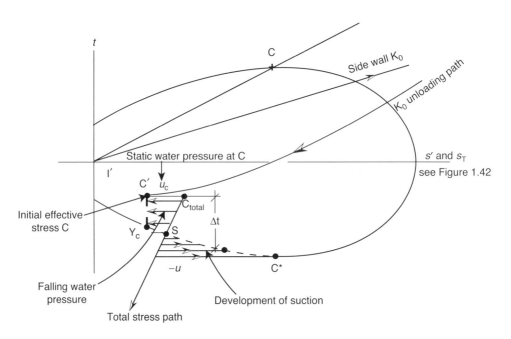

Figure 2.41 Stress path for a crown element.

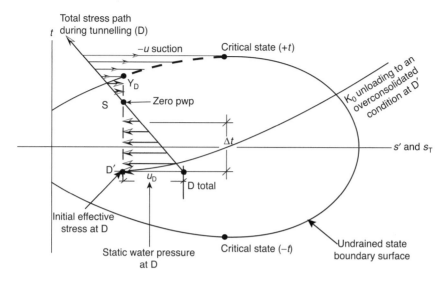

Figure 2.42 Stress path for side element on the horizontal centre line.

Units C_2 and D_2 will be lost during excavation but will give useful data forward of the face, particularly if the face is stabilised just before the instrument section due to a planned break (e.g. weekends, public holidays, etc). It may be possible to reconnect D_2 immediately after excavation.

The initial conditions in the ground need to be known, the vertical effective stress can be assumed to be $\sigma'_z = $ depth, x the submerged density of the soil γ_{sub} (approximately $d \times 10$ kN/m^2); the horizontal stress σ'_h in over-consolidated clays can range from approximately $0.8 \times \sigma'_z$ to $2.5 \times \sigma'_z$. The horizontal stress can be estimated based on prior experience, or preferably measured with a Maynard pressure meter or self-boring pressure meter (sometimes referred to as a Camkometer). Remember the static water pressure will be the initial datum.

Figures 2.38 and 2.39 show the soil elements C (crown piezometer B_2 in Figure 2.37) and D (on the centre line, piezometer C_3 in Figure 2.37); Figures 2.40 and 2.41 show the two total stress paths during geological loading and subsequent K_0 unloading to the over-consolidated stress states D′ and C′. The static water pressure gives total stresses at D and C respectively. As the excavation proceeds, the total radial stresses on the tunnel wall fall to zero (atmospheric pressure) while the circumferential stresses rise due to arching in the close proximity of the unsupported tunnel wall.

Note: In very simple terms the rise in the vertical total stress at the centreline level due to arching over the tunnel will be less than the corresponding change in the horizontal total component at the tunnel crown (and indeed the invert) due to the over-consolidation stress difference.

For an over-consolidated clay the natural soil has been overloaded under K_0 conditions to some high value of s'. Then it has been naturally unloaded by erosion etc., still under K_0 conditions, to its current *in situ* stress state with lateral effective stress, say 1.5 times higher than the vertical effective stress. The pore water pressure being the static ground water head.

This condition is shown by points C_{total} and C' and D_{total} and D' in Figure 2.40. Note t is consequently *negative*. As the tunnel is excavated, the total radial stresses on the tunnel wall fall to zero. At the crown this means that the vertical total stress falls to zero, while the horizontal stress arches round the excavation, raising the total horizontal stress in element C. Note that overall the effect of this is a fall in s_{total} while t rises very rapidly, it rises in the same negative direction ($\sigma_h \gg \sigma_z$).

At the centre line the horizontal total stress falls to zero, while the vertical stress arches round the tunnel, raising the vertical total stress in element D. There is a greater fall in s_{total} on the centre line as loss of σ_h is large and the arching increase is smaller than at the crown (the lateral effective stresses were $1.5 \times \sigma_z'$). *But* the change in t is in completely the opposite direction as in this case $\sigma_{h(total)}$ falls and $\sigma_{z(total)}$ rises.

The two *total* stress paths starting from C_{total} and D_{total} are shown in Figures 2.41 and 2.42, respectively. Their effective stress paths start at C' and D' and follow the suggested vertical path to yield, and then on the yield surface towards critical state undrained failure.

The crown element (Figure 2.41) is initially very close to the yield surface and the rapid increase in negative t means the effective stress path reaches Y_c very quickly and yields plastically. By S the pore pressures are becoming negative (suction). (Note: the invert behaviour is similar but starts from a higher s_{total} and S' value and hence the t change to the yield point is greater *but* the path from its yield to C* will show less development of suction.)

The change Δt in Figure 2.41 shows the element approaching critical state. However, the centre line element D has a total stress path (Figure 2.42) in which s_{total} falls slightly more rapidly than at the crown, *but* t changes rapidly in the positive direction, although in this case the value Δt still remains rather longer in the elastic regime until Y_D, still with the development of suction pore pressure.

The resulting behaviour is that the soil behaves nearly elastically as the effective stress path crosses within the boundary surfaces, the pore water pressure reduces steadily, becoming negative at S until the yield surface is reached at Y_D when the rate of suction pore pressure rises significantly.

Figure 2.43 suggests the general shape of the pore water pressure distribution around the tunnel with negative pore pressures developed in the shaded areas. If significant plastic shear deformation develops, it tends to become non-uniform and favours mechanistic yield lines. The result is additional suctions developing on specific shear planes, and if there is sufficient pore pressure data the distribution may resemble Figure 2.44.

It must be appreciated that, unless there is a comprehensive array of pore water pressure units, this detail may not be capable of resolution. Marked *changes* in the *rate* of development of falling pore water pressures, prior to lining completion, should always be subject to immediate review.

Immediately excavation of the tunnel has ceased in the vicinity of the instrumented section, the pore water pressures will rise back towards the original static water level, unless a tunnel is acting as a permanent drain. The pore water pressure response with time should look similar to Figure 2.45.

In major tunnelling works in soils for which there is not considerable known prior experience, a trial instrument section incorporating comprehensive vertical settlement and horizontal displacements and pore water pressure units is recommended.

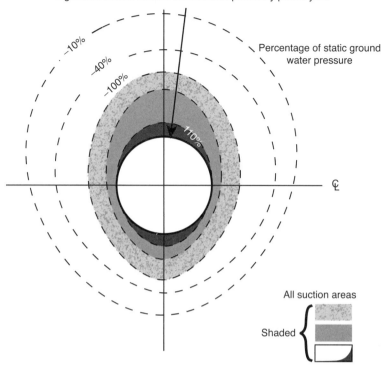

Significant suction zone in crown accompanied by plastic yield

Percentage of static ground water pressure

−10%
−40%
−100%
110%

C̶L̶

All suction areas

Shaded

Figure 2.43 Pore pressure suction distribution round a tunnel construction.

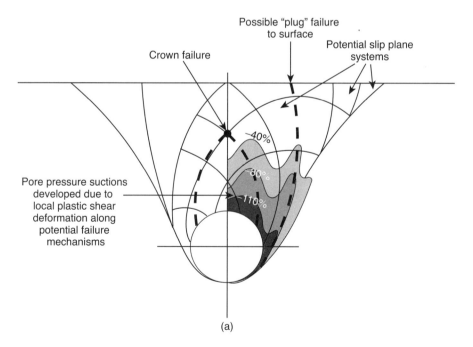

Possible "plug" failure to surface

Crown failure

Potential slip plane systems

−40%

−80%

−110%

Pore pressure suctions developed due to local plastic shear deformation along potential failure mechanisms

(a)

Figure 2.44 Contours of pore water pressure changes when Δt exceeds yield and rupture mechanisms begin to develop.

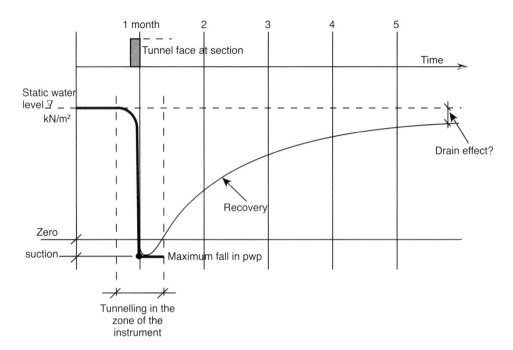

Figure 2.45 Changes in pore water pressure with time assuming rapid continuous tunnelling.

For ongoing construction a limited number of key locations can be identified by the trial, and during normal construction these are monitored against the performance in the trial section.

For tunnelling in materials other than over-consolidated clays, pore water pressures should still be monitored, but mainly to identify the development of wider falling ground water levels, which can identify drainage flow and warn against serious settlement resulting from increased effective stresses under adjacent buildings. With compressed air pressure tunnelling and pressurised shields, local pore water pressure measurement may help identify potential blow-outs, but these occur very abruptly and the warning time may well be insufficient.

Chapter 3

Displacements and global deformations

As indicated in the Introduction and Chapters 1 and 2, groups of data are key to theoretical analysis, and if obtained using instrumentation are key to the assessment of geotechnical and structural problems:

a) Stresses, both normal and shear; and
b) Strains or deformations.[1]

Instrumentation usually measures loads or displacements.

General

Stresses in soils are particularly difficult to determine, as in reality they consist of the summation of a vast number of point contact forces from each soil grain onto an area of a measuring device.

The contact face cannot replicate the soil structure and so the distribution of contacts will *not* be representative of the soil structure, nor will the body of the transducer mimic the deformation behaviour of the volume of soil it has replaced. The area of a pressure sensor will be relatively small and the unit will either attract load or under-read load, depending on the relationship between the unit's elastic characteristics and those of the soil. Further, almost all planes within a homogeneous soil mass experience both normal and shear stresses. No currently available commercial pressure cell reads shear stress and normal stress on the same surface. Special cells capable of measuring both normal and shear were made by Cambridge University in the 1960s and 1970s for contact stresses on retaining walls and other structural faces but are not commercially available. Direct measurement of stress in soil is therefore inherently unreliable.

The only stress in geotechnical engineering that is reliable is the 'pore water pressure' and its changes (u and Δu). This stress requires certain external conditions to be met, see Chapter 2:

a) The ground is fully saturated.
b) There is a complete and saturated contact between the ground water and the de-aired fluid inside the pore water pressure transducer.

---

1. Fibre optic sensors can now measure continuous stress along individual fibre optic lines.

c) A comprehensive understanding of the soil's behaviour under the applied engineering loading conditions is needed to correctly interpret and gain the most from the pore water pressure responses as indicated in the previous chapter. The value of u may also be predicted by numerical analysis.

Strains and deformations

Displacements are real. Displacements clearly identify physical changes which can be monitored by direct measuring techniques (from survey systems to micrometers). All systems require two basic references:

a) A fixed reference location at some stable point, a datum in x, y and z.
b) Orientation axes for the x, y and z direction.

All displacements are movements relative to the datum. Strains are differences in displacements over a local known original gauge length, i.e. $\Delta \iota / L_o$ and all strain instruments work on this principle.

The choice of gauge length is a product variable. It can range from less than a centimetre for electrical strain gauges and crack gauges, to several metres for a beam sensor unit. The change relative to the gauge length can be measured by numerous physical techniques, from resistance or capacitance change to one of the most useful, the variation in tilt angle with respect to a gravity reference $\dot{g}$. Gravity reference $\dot{g}$ instruments when set in a vertical line are used to determine one or both horizontal displacements, i.e. Δx and Δy, and hence require a horizontal reference direction (either magnetic North, site grid North or a local structural axis). These units are Inclinometers and, when used in a horizontal line, they directly determine δz and are termed horizontal profile gauges (HPGs). Settlements are a direct measurement of displacement relative to some datum, a benchmark to the ground level at a settlement point, or the change of level compared to a deep bottom of borehole reference, which is assumed not to have moved.

The above has outlined the basis of load and displacement measurement. The following chapters will deal with these two areas in more detail, starting with displacements as these are the more easily conceived system.

Vertical displacements within structures and soils

Within the field of displacements, the use of horizontal profile gauges and in-place inclinometers (IPIs) for either vertical or horizontal displacements will be covered in detail later (page 61 onwards), followed by combined systems (IPI sensors and horizontal profile gauges). There are, however, a number of *individual* instruments that will record either absolute level against a benchmark, or simply determine the relative level of one individual target point against adjacent target points. This is analogous to intermediate point levelling by ground survey (a subject that will also be included within the discussion of total station survey in Chapter 5).

The method of locating the individual target points and relating their level at the time of measurement can be achieved in a number of simple to very sophisticated ways.

Hydraulics

The easiest and probably most reliable method of levelling was invented in pre-Roman times and uses the fact that water will always find its own level (even compensating for the earth's curvature) provided the atmospheric pressure at both ends is the same, so using a pipe full of water connected between a datum and a target point will always provide a level at the target point.[2] The simplest system is the *overflow cell* and the datum house. The target levels concerned are within a narrow level band of the gauge house datum (± 1 m) and they are not anticipated to fall below the gauge house floor. Figure 3.1 shows the system diagrammatically.

Water is passed from a container in the gauge house to the unit and overflows into the unit from pipe A and returns to the gauge house via pipe B to a second container H below (Figure 3.1). When water is flowing out of B, the two containers in the gauge house are lowered until the two levels show no flow. The level in the gauge house is then the level of the unit. The drain is then opened to clear the device. A series of valves enable a number of target units to be levelled from one gauge house.

This system is basically very simple and reliable, but is limited to a specific horizon and is appropriate for new construction above ground where the gauge house

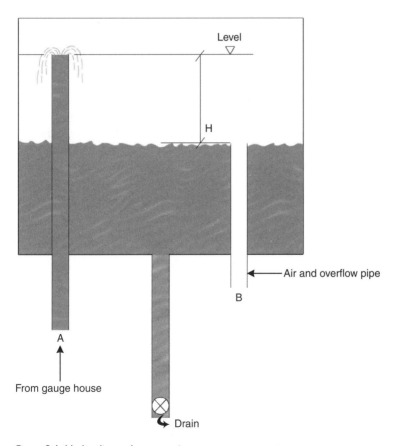

Figure 3.1 Hydraulic settlement unit.

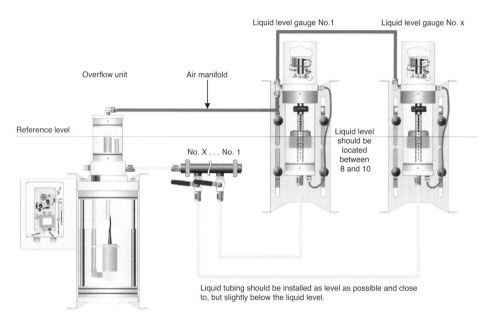

Figure 3.2 Comprehensive automatic hydraulic settlement system used inside structures (also known as a 'water level' system).

can be easily located on the same horizon, earth dams being the primary use where there may be five levels and gauge houses to monitor the dam settlement. The air pressure in the overflow unit must be the same as the air pressure at the gauge house, the drain and pipe B ensure this. Some care must be given to the water and the pipes, which are small-diameter nylon and can be very long, and are prone to organic growth which can block the pipes. Antifreeze and potassium permanganate are often added to mitigate problems. A sophisticated version for inside structures is shown in Figure 3.2.

Piezometer settlement cells

It is somewhat restrictive to be limited to a single horizon above ground level. Water is, however, a continuum and a fluid, which means its pressure will be a direct measure of the height of a column of water provided the temperature and pressure remain constant. The temperature of water buried more than 2 m below ground level remains virtually constant (at 12°C). If, therefore, a datum water level is maintained in a gauge house and the target cell set any distance below this settles, then the height of the water column h above the target cell will change. The water pressure in the target cell will increase and this can be measured by a piezometer (Figures 2.15–2.17). Almost all current piezometers are sealed and calibrated against a vacuum reference on the sealed side of the diaphragm. The pressure in the cell is therefore a total absolute pressure against vacuum zero and the atmospheric component is included in the reading, Figure 3.3 shows the scheme.

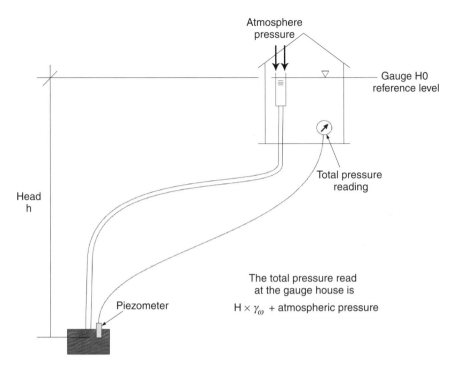

Figure 3.3 Piezometric settlement system.

The atmospheric pressure can vary considerably from $+950$ to $1050\,\text{mbar}$ ($100\,\text{mbar}$ $\approx 1\,\text{m}$ of water), which represents approximately $\pm 500\,\text{mm}$ of water head. Correction must be made for this and a Druck or similar pressure cell must be incorporated in the system at the reference level.

As with the previous instruments, the tube must be guaranteed to be unblocked and full of water; a recirculation system is always incorporated to ensure this, and prior to taking a set of readings the system must be carefully flushed.

Air bubbles in small pipes suffer seriously from the meniscus drag of air bubbles. A $3\,\text{mm}$ inner diameter tube will give a meniscus rise of between $1\,\text{mm}$ and $3\,\text{mm}$, depending on the inner surface of the tube. Each bubble when pushed requires two meniscus forces to be overcome. Each air bubble could represent a head loss of $6\,\text{mm}$ in real head of water plus the missing length of water l_b (see Figure 3.4). Water pressure $= (h_1 - 6\,\text{mm} + h_2)\,\text{m}$ of water, *but* the real head should be $h_1 + h_2 + l_b$, so the pressure loss for each bubble $= (l_b + 6\,\text{mm})\,\text{m}$ water.

It is therefore essential to completely de-air the water in the tubing before taking a reading.

These units are sized to be installed either within pockets in newly placed material or grouted at the base of a borehole, the pipework and cable from the piezometer being carefully protected in a trench back to a gauge house and datalogger. Again, subject to the limitations of pipework length, multiple target units can be de-aired and read from one location, not limited in this instance, to one horizon.

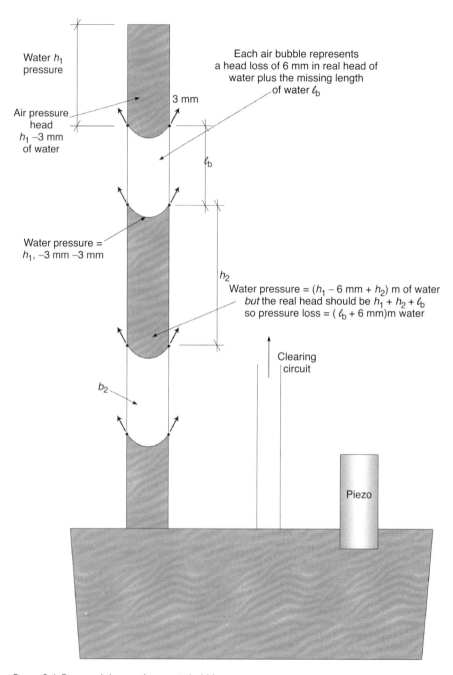

Figure 3.4 Pipework losses due to air bubbles.

Displacement of structures

The exterior of surface structures can be monitored to an acceptable accuracy by modern surveying techniques using total stations, laser scanning technology and photogrammetry (see Chapter 5). Internally and in confined basements these techniques may not be acceptable. Beam chains, i.e. a continuous series of electronic spirit levels, have been used for many years. The system is relatively simple, a series of reference pins are installed near horizontally at 1–3 m centres along walls or vertically up a wall or column. Stiff, light beams (typically 38 mm square section aluminium) are mounted between the reference pins. One end has a fixed bearing carried on one pin, the other end is a sliding bearing on the second reference pin.

The beams should be continuous. A levelling unit is installed within or on the beam containing an electrolevel or microelectromechanical systems (MEMS) tiltmeter. The data outputted by the system are the small angle changes $\Delta\theta$ applied to the horizontal distance between the reference pins, and gives the relative vertical displacement ΔV at the second pin, see Figure 3.5.

The principal of an electrolevel is effectively that of a spirit level, in which the position of the bubble can be measured as indicated diagrammatically in Figure 3.6 and assessed by a standard half-bridge (AC) circuit as shown in Figure 3.7. The output to tilt angle is obtained by direct calibration.

The overall performance of electrolevels has been the subject of much debate over the years and depends on the constituent parts of the unit and its calibration, most importantly:

- The actual electrolevel vial used.
- The way in which the vial is mounted.
- The method of excitation and subsequent signal conditioning and analogue to digital conversion.
- The calibration applied (both for tilt and, if needed, temperature).

Many varied types of electrolevel vial are manufactured, some specifically designed for a single use (e.g. gravity reference zero). The ranges available are from approximately 1° to 360° with, in general, the greater the range, the lower the repeatability and accuracy.

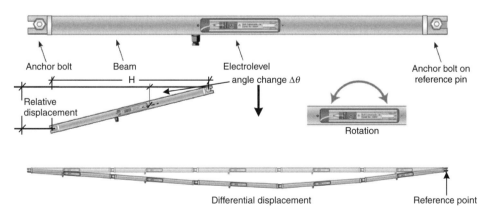

Figure 3.5 Line of electrolevel beams.

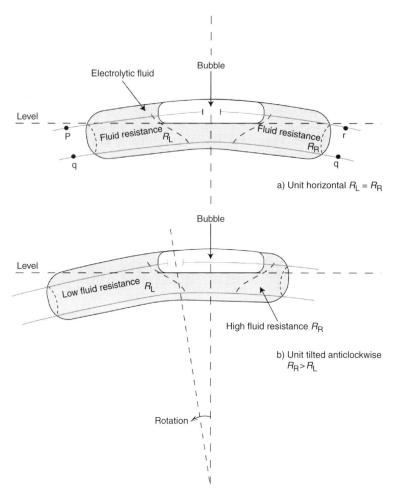

Figure 3.6 Principle of resistance changes in an electrolevel.

The most common type used in geotechnical and structural monitoring has a $3°$ range and a reasonable linear resolution of $\pm0.75°$. As a non-linear device it requires either a third- or fifth-order polynomial to gain the accuracy required (see Figure 3.8).

The method used to secure the vial into the mounting system is fundamental to the performance of the instrument. Non-symmetrical mounting, non-axial orientation and non-uniform temperature distribution may result in errors. To overcome these potential problems, carefully designed mountings are produced, and specialised mounting materials, such as ceramics and low-moisture-absorption resins, are commonly used, with the aim of providing a reliable and stable coupling of the vial to its carriers. A similar approach must be adopted in the mounting of the vial carrier into the beam.

Incorrect excitation of the electrolytic fluid contained within the vial will result in electrochemical effects that will change the properties of the electrolyte, and in

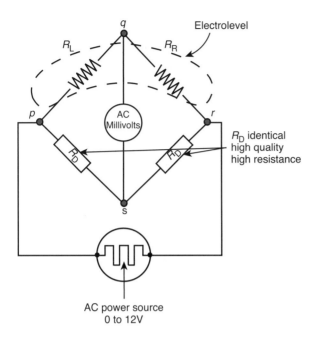

Figure 3.7 Wheatstone bridge for electrolevels.

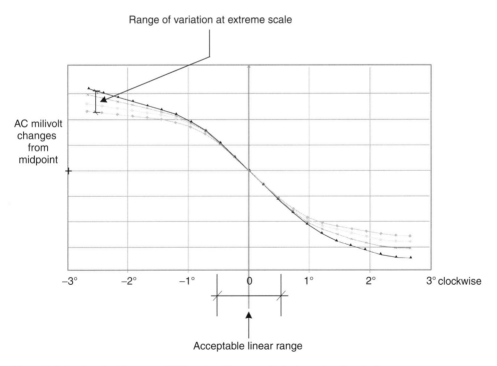

Figure 3.8 Fredericks Company 0711 series (five samples) electrolevel typical output.

particular its sensitivity, rendering any calibration invalid. Further effects of incorrect excitation include sensor drift and an increased sensitivity to temperature.

It is necessary to accurately determine the sensitivity of the electrolevel and in particular to characterise its non-linear behaviour. Each vial has slightly different physical properties and, as such, differing sensitivities and non-linearities between any two samples (see Figure 3.8). Thus a multi-point calibration is required throughout the working range of every sensor. Thankfully this process can be automated and results in either a third- or fifth-order polynomial for correction of raw data to the desired engineering units (most commonly mm/m). It is possible to use a linear calibration over the centre section of the vial's output curve, which can yield extremely accurate results. If this limited range is to be used, mounting and zeroing must be done with great care on-site. The working range can be as low as ±4 mm/m (although this is probably sufficient for structural monitoring as it represents a deformation of 1 in 250).

One of the commonest electrolevel vials in use is the Fredericks Company 0711 series, which has a notional full scale range of ±3°. The calibration over the full range as discussed above is non-linear, and differs between individual samples, as shown in Figure 2.8.

Incorrect use of electrolevel beam systems

There are two common uses of beam technology, which are of very limited analytical use. The first is a series of non-aligned units as typified in Figure 3.9. Data will be shown for a typical temperature change.

Figures 3.10 and 3.11 show the displacement for unit number six, which as indicated earlier (in Figure 3.5) will provide only the horizontal displacement (Δt_{90}) but not

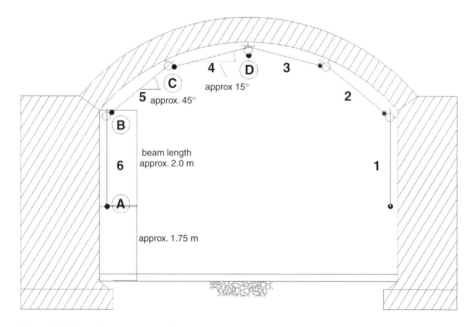

Figure 3.9 Tunnel section: non-alignment beam system.

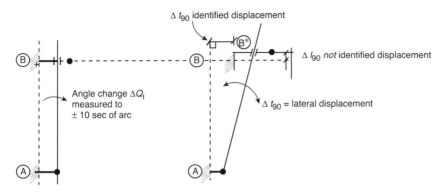

Figure 3.10 Limitation, unmeasured vertical displacement measurement on a vertical bar unit.

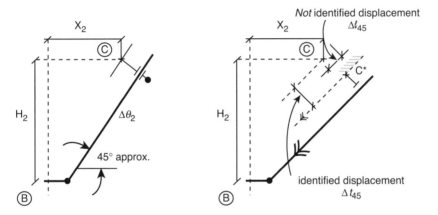

Figure 3.11 Limitation, unmeasured axial displacement measurement in an angled bar unit.

the vertical value Δl_{90} If this is a monolithic structure then Δl_{90} may be very small and thus not of concern. However, for inclined units 5, 4, etc., only the transverse displacement Δt_{45} and Δt_{15} etc. will be determined; in every case the Δl_{45}, etc., will remain unknown. Figure 3.12 shows the development of the unknown location of pin C. In Figure 3.13, the known displacements of the arch between B and F are shown. In Figure 3.14, the possible real positions are shown using the best estimates of the unknown Δl displacements (dashed lines).

It was known from survey that point F did not settle at all and moved away from B by only 0.25 mm. The discrepancy must be compensated by the length changes in the arch line. In Figure 3.14, the best fit has been drawn with the unknown measurement shown as the dashed components. The in-arch components are subjected to expansion because of the temperature changes and compression due to increased stress in the arch. Assuming uniform temperature rises the variation in length changes from 1.34mm (BC), 0.96mm (CD), 0.8mm (DE) and 0.58mm (EF), suggests increasing compression towards arch seat F, presumably as a result of thrust against the overburden, which pushed the abutment out the 0.25 mm.

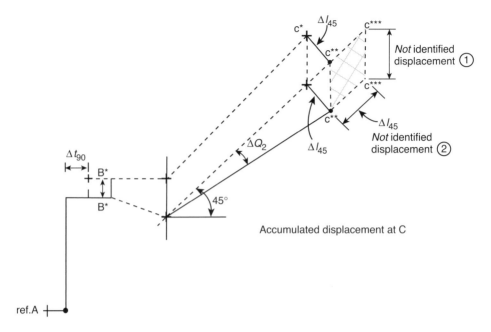

Figure 3.12 The influence of unknown displacements on the real position of pin C.

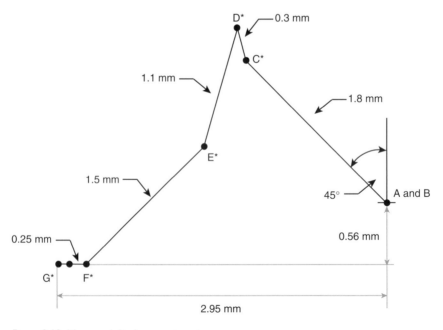

Figure 3.13 Measured displacement vectors.

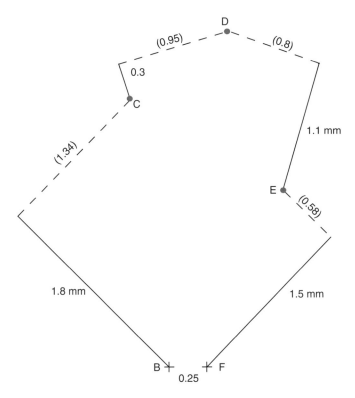

Figure 3.14 Best-fit estimate of in-line length changes to match surveyed displacements of F.

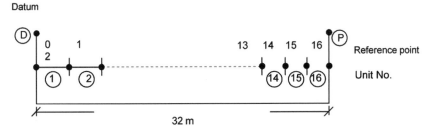

Figure 3.15 Location of reference pins and beams 1 to 16.

The second example is the simplification of a continuous beam chain by adopting a tilt meter approach, installing independent tilt meters or beams on a 'hit-one–miss-one' basis, as illustrated in Figure 3.15.

In this case, the 16 beams and electrolevel units were installed from D to P and observed during a deep excavation at P. The full displacement data on completion produced the settlement profile shown in Figure 3.16, with a maximum value at P of 51 mm and the corresponding critical angular distortions shown in the table in Figure 3.16.

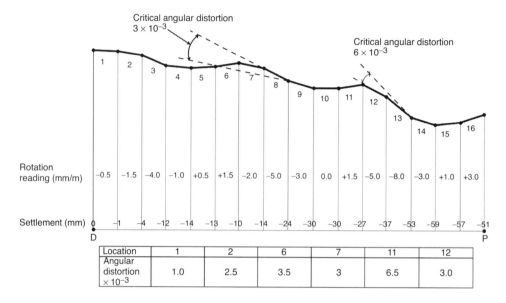

Figure 3.16 Full settlement data for a continuous chain.

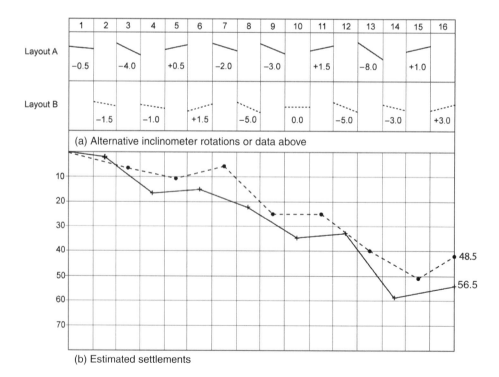

Figure 3.17 Estimated settlements based on alternative tilt meter layouts.

Assuming the two 'hit-one–miss-one' tilt meter layouts had been adopted, i.e. beam units 1, 3, 5, etc., or 2, 4, 6, etc., the extrapolated displacement curves would have been as shown in Figure 3.17, giving displacements at P of 56.5 and 48.5 mm, respectively, and completely misleading angular distortions (see Boscardin and Cording, 1989).

Horizontal displacements in structures

In stiff structures, the accuracy and really long-term stability of inclinometer systems may not be acceptable. This section focuses on horizontal displacements in masonry and concrete structures, in particular dam structures – both gravity and arch designs where measurements due to seasonal filling may last for decades.

Vertical displacements are usually not a serious consideration in tall buildings (the Tower of Pisa excepted) or concrete dam structures; however, the horizontal movements due to water impounding or earthquake loading need to be known and checked against theoretical predictions. The key interest is the relative horizontal displacement of the structure at various levels with respect to the stable rock foundation. One of the well-established instruments for this is the standard (hanging) or inverted pendulum. Figure 3.18 shows the elevation and plan of this system diagrammatically.

For the inverted pendulum, a thin wire is anchored in the rock foundation and is supported at or near the crest by a float within a special bath of oil or water. The wire is tensioned by partly submerging the float, so that the excess buoyancy has to be carried by the wire. The buoyancy force will always be vertical, so if the concrete structure moves sideways the wire acts as a vertical reference line. The lateral displacement of the structure can be measured by any form of non-contacting displacement transducer system; often two non-contacting units will be installed set at 90° to one another, one is shown in Figure 3.19 using an optical system.

Displacements are anticipated to be small compared to those experienced in soil structures, and the use of precision micrometer- or laser-based readout systems with a range of between 40 mm and 100 mm to monitor the position of the wire is usual. They can be installed in galleries at any level within the dam, the measuring unit being optical if read at extended intervals, or a laser-beam sensor if automatic readings are required. A very recent modification requiring no moving micrometer or other parts is the use of two charged coupled device (CCD) pixel arrays similar to those used in digital cameras, with a parallel light source opposite. The shadow cast by the wire on the pixel strips gives an immediate location in x and y.

The standard vertical pendulum is top mounted (anchored) with a heavy mass at the base; reading systems are identical to the inverted system but act in reverse. The one disadvantage of this unit is that, as with all long pendulums, the earth and moon's gravitational pull will tend to result in an automatic swing which precesses. A viscous damping oil bath has to be installed at the base to prevent this and care must be taken not to disturb the system when reading. For these reasons, the inverted pendulum, which does not suffer from these effects, is now the more commonly used.

Vertical displacements in soils by mechanical means

Survey techniques use classic optical instruments to record levels during and after construction, setting studs in surface concrete blocks and performing standard optical

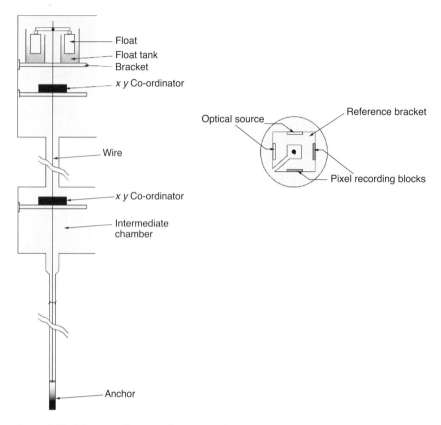

Figure 3.18 Schematic diagram of an inverted pendulum system.

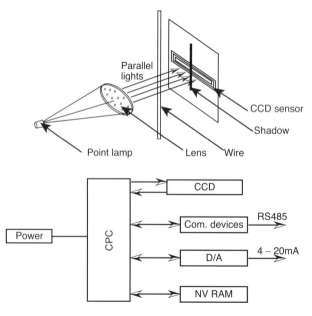

Figure 3.19 Principle of the optical pendulum reader system.

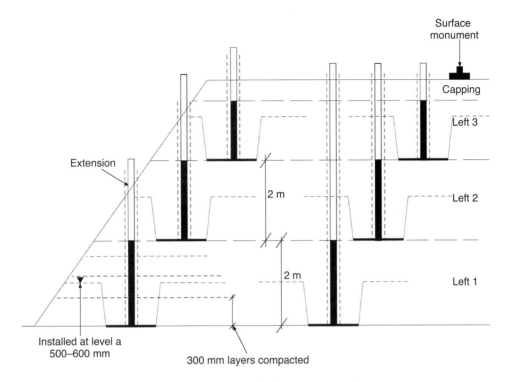

Figure 3.20 Use of plate settlement units in an embankment construction.

levelling circuits from a benchmark, at regular intervals. This data technique is normally required to close within ±3 mm but such stud levels are often taken not as *change points*, but as intermediate sights, and the accuracy is more probably ±5 mm. The concept can be extended for above-ground construction by installing a horizontal steel plate 0.5–1 m below the current fill level onto which is attached a vertical length of steel tubing (normally 2 m long). The plate and tube are carefully backfilled using a plastic cover pipe to isolate the steel tube from the surrounding fill. The top of the steel tube can be levelled as filling continues and additional lengths of steel tube and plastic sleeve added as required. Figure 3.20 shows an idealised system.

This system is crude and simple; it is also very prone to serious damage from construction equipment (various methods of protection are usually fabricated from materials on site such as oil drums). If the site personnel can ensure a rigorous protection routine, the system is reliable.

Micrometer settlement measurements (the rod extensometer)

A highly sophisticated variation of the crude principle outlined above is to anchor a number of precision fibreglass or steel (mild or stainless) rods within a borehole at key levels in the foundation. The rod is again surrounded by a sleeve to ensure the rods are

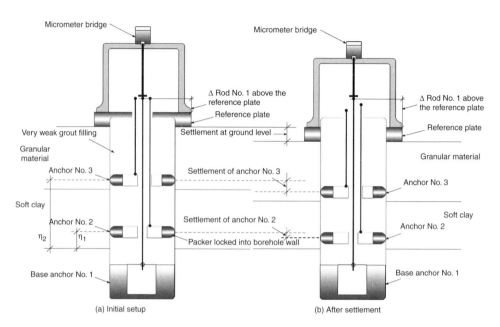

Figure 3.21 Rod extensometer settlement measurement.

independent of the grout-filled borehole. The key levels are, for instance at two metre centres down a borehole as suggested in Figures 3.30 and 3.62 (later) (up to eight), or alternatively at key strata changes. The rods extend from their fixed anchor locations to the borehole top where they are located in a reference plate. If there are differential settlement values experienced down the borehole, then the relative settlements will be replicated by the tops of individual rods. A standard 0.01 mm micrometer can be mounted on a 'bridge' unit and be used to assess the differences by measuring the rod ends against the reference plate. One rod should be anchored in a stable base strata which is then used as a datum. Alternatively, electrical displacement transducers can be fitted and the system read remotely.

With this arrangement one borehole can accommodate up to eight measuring extensometer rods, the system is shown diagrammatically in Figure 3.21.

The individual rods can be located at their lower ends by a number of methods from an individual high-pressure cement-grouted ring packer with the remainder of the borehole filled with a weak bentonite/cement as idealised in Figure 3.21(a), or by any mechanical anchor block in a continuously grouted hole, see Figure 3.22, which shows the details of the unit and its component parts.

The original micrometer units were read manually. If an automated system is available the measuring head can be replaced by a set of electrical transducers (e.g. linear potentiometers or vibrating wire displacement transducers). The data from each rod and displacement transducer can be digitised and transmitted to, or interrogated from, a remote location.

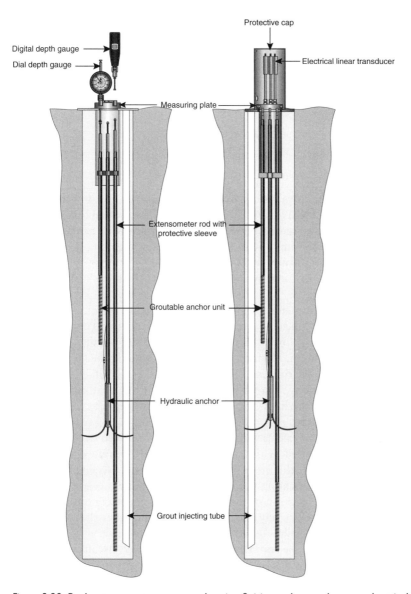

Digital depth gauge

Dial depth gauge

Protective cap

Electrical linear transducer

Measuring plate

Extensometer rod with protective sleeve

Groutable anchor unit

Hydraulic anchor

Grout injecting tube

Figure 3.22 Rod extensometer systems showing fixities and manual versus electrical measurement.

Magnetic plates and reed switch sensors (the magnetic extensometer)

As indicated later in this chapter (page 60 onwards) in the discussion of inclinometer measurement, considerable benefit can be achieved by combining the vertical and lateral displacement measurements to provide a full Δx–Δy–Δz picture of an area. The ideal being to obtain strings of data on the same inclinometer line. The most expensive element of any instrumentation system which includes inclinometers is the

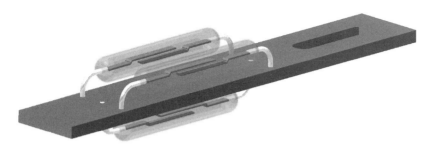

Figure 3.23 Reed switch unit.

formation of the boreholes, the installation of the inclinometer casing and the careful grouting of the hole/casing. An ideal casing, 70 mm in diameter is available for any type of probe down to full depth; however, no borehole this diameter can be bored to depth. Standard holes are typically 100 mm, 150 mm and 200 mm in diameter. Outside this tube is a grouted annulus from 100 mm to 200 mm in diameter 'intimately' connected to the soil. How can this zone of the construction which is 30 mm to 130 mm wider than the casing be used?

Remembering that the inclinometer casing is fabricated from plastic, a magnetic ring can be located in this annulus. Magnetically sensitive electrical reed switches (Figure 3.23) are available, which close to form an electrical circuit within a quite weak magnetic field. Closing a circuit can illuminate an LED and sound a beeper/buzzer. If the magnetic field produced by the ring can be placed at a fixed position in the ground, its depth can be located and read by the reed switch passing through the magnetic field as shown in Figure 3.24 (where the heavy black lines show the magnetic field strength capable of closing the reed switch).

A reed switch passed down the casing will close when it passes into a magnetic field of sufficient strength and will open again as it passes out. If the two depths to the on and off positions are noted as the probe is lowered, and again when the probe is raised, four values are obtained locating the band of sufficient magnetic strength and then the mean will provide the location of the horizontal magnetic centreline.

The key requirement is to effectively fix the magnetic ring into the wall of the borehole. An early device was a plate containing three pointed pistons set at 120°; air pressure forced these out into the soil. Unfortunately, all three pistons rarely operated identically and often only two would fix. The much simpler mechanical device shown in Figure 3.26(b) (later) is now most commonly used. It consists of three double spring-loaded arms with end points that dig into the borehole wall when released.

Figure 3.25 shows diagrammatically the installation sequence, the key action being the removal of the locking pin as the drill casing is extracted and the grouting proceeds. Figure 3.26(a) shows the real arrangement of three units and two plates in overlying fill.

In Figures 3.51 to 3.53 (later), horizontal profile gauge systems will be shown with magnetic plates placed in the instrument trenches surrounding the casing. These plates are, of course, much simpler to install in open trenches and Figure 3.27 shows a typical plate used on a vertical axis. These are commonly installed at 2 m to 5 m intervals.

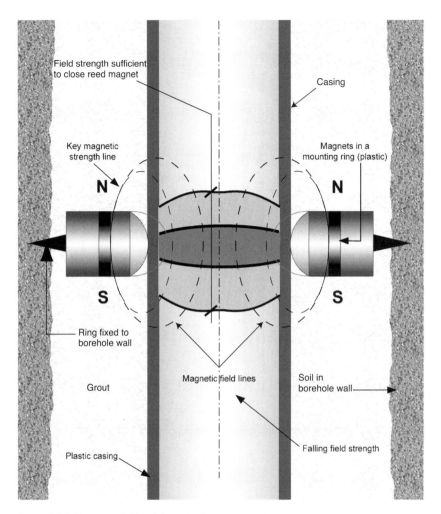

Figure 3.24 Magnetic field inside an inclinometer casing.

Deformations in soils using inclinometers and hydrostatic profile gauges

The fundamental of all these systems is an average angle change over L_0, i.e.

$$\frac{\Delta z}{L_0} \text{ or } \frac{\Delta x}{L_0} \text{ etc.} = \tan \Delta \theta = \Delta \theta \text{ (in radian units when } \theta \text{ is small)}$$

A chain of units with constant or various L_0, each measuring individual changes will, by the addition of each component, provide the profile of one line on or within a soil/structure problem. Geotechnical projects involve a continuous material 'half-space' many metres in area and often 30 or more metres in depth. The use of several

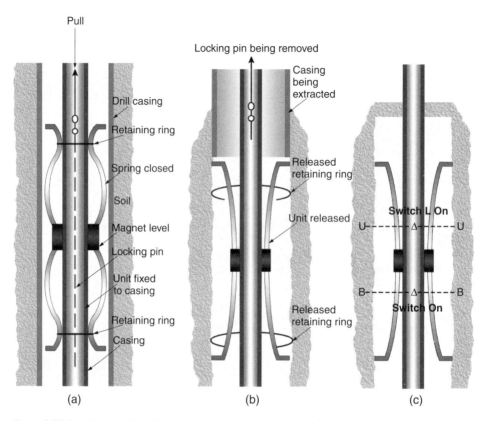

Figure 3.25 Installation of a spider magnet system outside the inclinometer casing.

such instrumentation lines can provide a comprehensive understanding of the response of the soil to the changes being carried out by the contractor.

A vertically aligned inclinometer utilises the gravity reference to determine the Δx and Δy profile. A horizontally aligned hydrostatic profile gauge (HPG) also uses the gravity reference and will determine δz. When using a vertically aligned inclinometer, the Δz changes can be measured by an independent level measurement system against an ordnance survey datum, at specific points within the soil mass by the various settlement systems available. These can be arranged at independent locations or in a static chain utilising the same location, even using the same casing as the vertical inclinometers.

Figures 3.28 and 3.29 show a single vertically aligned location together with the locations of basic vertical settlement points. Figure 3.29 shows diagrammatically the two key output diagrams: accumulated displacements versus depth, and slope change versus depth; the settlement vectors, e.g. Δ_{z_5}, being superimposed.

Before discussing data recovery and the interpretation of the data provided by an inclinometer and combined inclinometer/settlement systems, the technical details and installation methods will be discussed.

(a)

Fill

Plate magnets in fill

Grouted borehole

Spider locators

Base reference ring
fixed to casing

(b)

Locking pin

Spring retaining
loop

Spider springs

Probe

Retaining
loop

Closed spider

Magnetic ring

Alternative
magnetic ring

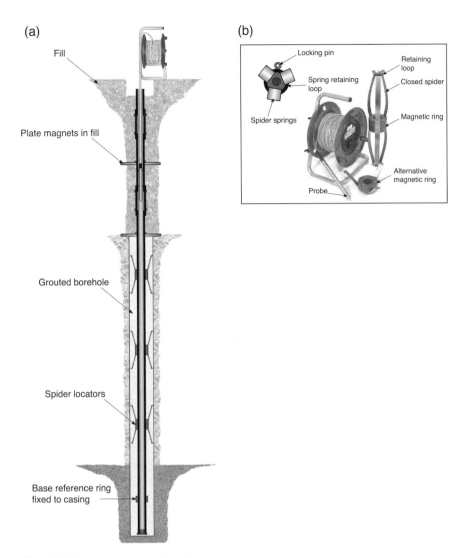

Figure 3.26 Spider magnet and probe.

Figure 3.27 Plate magnet.

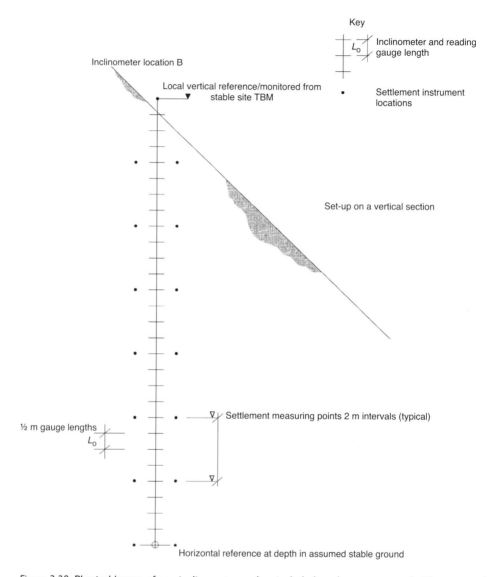

Key

Inclinometer location B

Local vertical reference/monitored from
stable site TBM

Inclinometer and reading
gauge length

Settlement instrument
locations

Set-up on a vertical section

½ m gauge lengths

L_0

Settlement measuring points 2 m intervals (typical)

Horizontal reference at depth in assumed stable ground

Figure 3.28 Physical layout of one inclinometer and an included settlement system (spider magnets).

Inclinometers: the technician

Having briefly reviewed the typical output of an inclinometer, this chapter will now examine the basic mechanical and electronic concepts and the reading equipment, followed by the installation techniques and the care required. The groups of civil engineering projects in which the measurement systems are useful will then be discussed. The techniques used for the installation of permanent facilities will be presented. The processing and presentation of basic data and its interpretation and significance will be discussed. Finally, advanced analysis of complex inclinometer settlement systems, such as those used for trial construction, will be demonstrated.

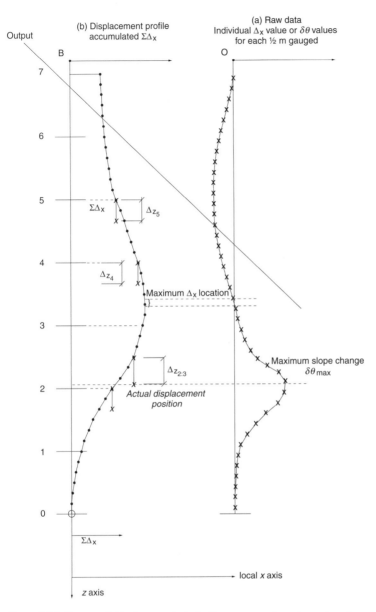

Figure 3.29 Data available from a single inclinometer and settlement system.

Inclination and the reading unit

Inclination is the angle off the vertical. Gravity is therefore the direction against which all information is referred. An inclinometer reading unit can determine changes of a near vertical location to better than 20 seconds of arc (1/180th of a degree).

However, the usual method of presenting data for engineering assessment is as a displacement equivalent, i.e. mm offset/metre of gauge length, usually quoting for an inclinometer unit between 0.05 and 0.1 mm/m.

A unit capable of measuring inclination will therefore incorporate a sensor in which the self-weight of some item resembling a stiff pendulum identifies the gravity direction with respect to the unit's physical axis. The relationship will be carefully factory zeroed before it is supplied and should be checked regularly when in use. A monitoring technique will then assess the angle of the gravity vector. Early examples were literally a sprung metal cantilever, strain-gauged to measure the bending moment at the root of the cantilever and calibrated. The output data were a simple analogue voltage change measured on a Wheatstone bridge circuit. This required a quality millivoltmeter and a power source. A more common and still currently available unit is a single weighted needle carried on precision bearings in the manner of chemical scales, in which the needle is returned to its instrument zero by a current in a coil, the amps required being the measure of the force required to reset zero. In effect, the reverse action of a moving coil milliammeter. Again, the output is an analogue voltage.

The most advanced current technique uses an accelerometer (based on capacitance changes) etched on a glass/ceramic chip, which is given in simplified form in Figure 3.30. Additional improvements can be gained by digitising the signal within the probe removing 'cable effects' experienced with analogue signals travelling to the readout.

The sensor must be carried in a stiff beam device with a fixed gauge length L_0. This in turn must register against a repeatable near-vertical reference line in the soil.

a)   The fixed gauge length carrier is formed by the body of the inclinometer probe.
b)   The repeatable reference line is provided by the casing.

Both need to be able to fully exploit the sensitivity and resolution of the sensor and will now be outlined separately.

The inclinometer body

Figure 3.31 shows a typical inclinometer body. It consists of a rigid central bar fixed to two sprung wheel carriers. The wheels are carefully machined to engage in grooves formed in the casing (see later), as in Figure 3.32.

The wheels are arranged as two sprung pairs set 500 mm apart and angled at approx 60° to the centreline of the body, as can be seen in Figure 3.33. When inserted in the casing, the spring no longer presses against the wheel stop but reacts against the two grooves located opposite each other in the casing, as in Figure 3.34.

This theoretically maintains good contact in both grooves and keeps the probe body exactly on the centreline of the casing. The spring loading arrangement compensates for any deformation in the circular section of the casing as shown (in exaggerated form) in Figure 3.35.

The upper end of the wheeled probe unit of the sensor housing is rigidly attached to a lifting cable; also connected at the top end of the probe are the electrical signal wires, all of which are factory sealed into the sensor housing to provide a waterproof system to the ground surface. The lifting cable is marked at 500 mm intervals with depth reference markers (Figure 3.36).

This unit can traverse a curved profile in standard (70 mm outer diameter (OD)) inclinometer casing, as indicated in Figure 3.37, giving a minimum radius of curvature

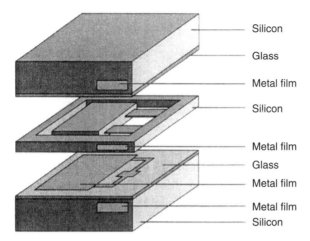

Figure 3.30 MEMS accelerometer (diagrammatic).

Figure 3.31 Complete inclinometer probe.

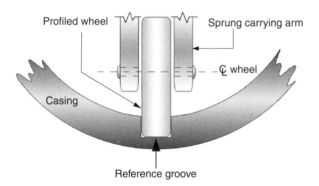

Figure 3.32 Inclinometer casing/wheel alignment.

of approximately 2.04 m, or a change in angle of 7.04° over the 500 mm gauge length. This is shown in diagrammatic form in Figure 3.38.

If with this degree of curvature the tube deforms to the elliptic shape shown in Figure 3.38, then the clearance may be further reduced. Smaller diameter tubes such as 48 mm OD and 58 mm OD casing will result in a major limitation to the curvature that the standard probe can traverse; this situation can be critical if any condition approaching failure develops, as the soil will fail over a very limited width of shear band across which all the gross displacements will be concentrated. Very tight local

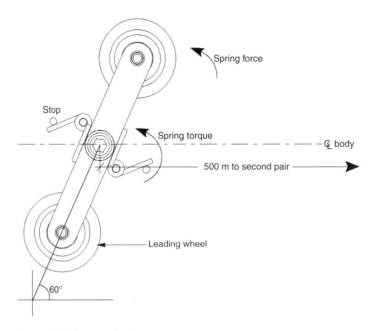

Figure 3.33 Sprung wheel action.

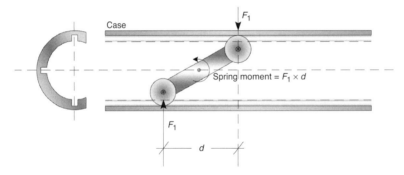

Figure 3.34 Forces on the sprung wheel pair.

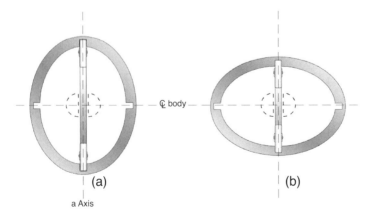

Figure 3.35 Sprung wheel action in distorted casing (diagrammatic).

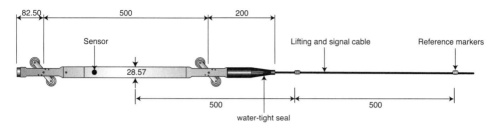

Figure 3.36 Dimensions of a typical inclinometer probe.

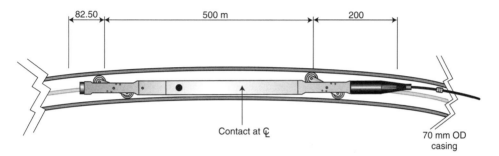

Figure 3.37 Inclinometer probe limitations within 70 mm outer diameter (OD) curved casing.

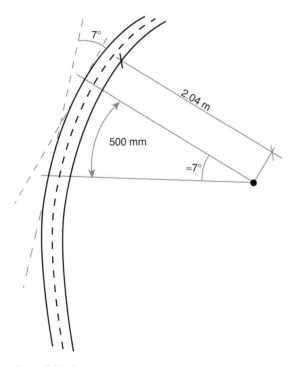

Figure 3.38 Exaggerated diagram of curvature limits.

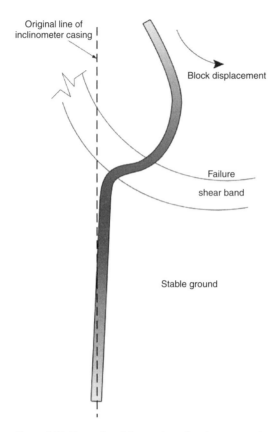

Figure 3.39 Excessive deformation of casing preventing use of a standard inclinometer.

radii may therefore occur together with casing distortion. A standard probe will not pass this point and may well become stuck in the casing, shown (exaggerated) in Figure 3.39.

Minimising instrument zero and small casing errors

As is apparent from previous figures, the casing is manufactured with four grooves set at 90° to each other (see later sections). The probe is carried on two pairs of sprung wheels located in the 180° grooves known as the A and B orientations or directions, as shown in Figure 3.40.

The +A alignment should always point in the *expected* direction of displacement and the *leading* wheel is always inserted in this groove to start a set of readings. During manufacture, great care must be taken to ensure that the centreline of the wheeled body of the probe and the gravity axis of the sensor are perfectly aligned, or are adjusted for when the instrument is truly vertical, and the wheels are in ideal contact with the grooves. In case this ideal situation does not exist, errors can be minimised by reversing the leading wheel by 180° from the +A to the −A direction and averaging the readings, see Figure 3.41.

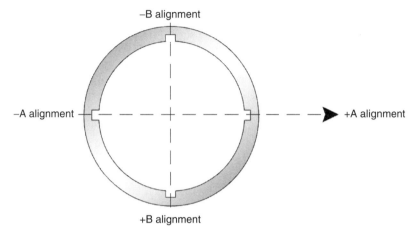

Figure 3.40 Inclinometer casing groove nomenclature.

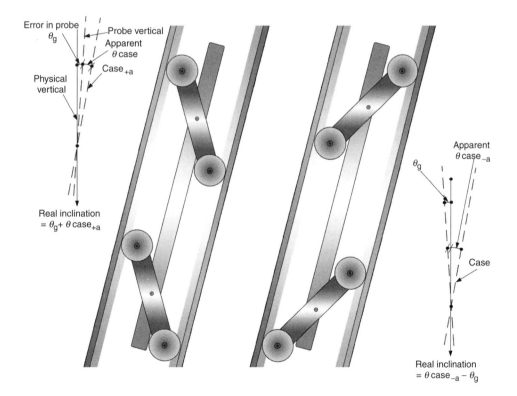

Figure 3.41 Elimination of sensor/body errors using a double traverse.

The average $= \frac{1}{2}(\theta_g + \theta \text{ case}_{+a} + \theta \text{ case}_{-a} - \theta_g) = (\theta \text{ case}_{+a} + \theta \text{ case}_{-a})\frac{1}{2} = $ real casing angle to true vertical.

And similarly for the B direction.

The previous sections outlined the basic mechanics of a quality inclinometer and its associated inclinometer casing. Later, an idealised situation maximising the combined output of a system of three or more vertical inclinometers and an associated settlement measuring system will be described. However, in many situations a simple monitoring system may be sufficient, and individual inclinometers with or without settlement measurements can be used to provide a basic but restricted output, shown in Figure 3.29.

Whatever level of sophistication is used, the output data will only be as good as the installation quality, accuracy, reliability and repeatability. The following sections will outline the methods and materials recommended to achieve the highest standards.

Fundamentals of inclinometer installation

The system is required to measure the horizontal displacements along a vertical axis (columns and walls) in engineering structures, in existing or in made ground (e.g. embankments). The four-groove specialised reference tube (inclinometer casing) outlined earlier must be installed on this vertical axis in such a way that it will always move exactly with the structure or the ground.

The horizontal orientation of the key measuring axis (the +A groove) must be known and should be within a consistent orientation. The measuring device (the inclinometer probe and readout, see previous section) will either traverse this vertical axis, taking angle measurements at specific intervals, or be installed as a permanent chain of tilt devices (in-place inclinometer sensors, see later).

In all cases displacements will be relative to a reference datum. If feasible, this reference datum will be at a depth in stable ground well below the zone expected to be influenced by any engineering work, or careful surveying must be used to reference the top of the hole to a site datum immediately prior to a set of readings. The static components of an inclinometer set-up are therefore:

a) A quality reference tube – the inclinometer casing.
b) An accurately drilled hole into which the casing must be carefully installed.
c) An installation technique which ensures that the casing is correctly orientated and is intimately connected to the walls of the drilled hole at all times and over the whole length.

As displacements are all relative to an initial reference data set, ideally, several sets of data should be obtained and averaged *prior* to any new work. However, in practice this often is not possible, as site access and contract agreements mean that instrumentation and engineering work often commence simultaneously. Under these circumstances a minimum of three reading sets should be taken as soon as the installation is complete and stabilised. The most representative set is then adopted as the datum. It should be noted that, even under natural conditions with no change in the environment except daily and annual weather changes, the upper levels of both structures and soils will move. The longer the period over which zero readings can be taken, the better the eventual interpretation and understanding of the significance of later data.

Figure 3.42 shows the three components in vertical section and in plan for a typical single installation.

The reference casing

Originally the casings were extruded aluminium sections and serious corrosion resulted in disintegration of the casing in some soils. Fibreglass was also available at that time but was very expensive, brittle and prone to corrosion. The advent of high-quality, high-stability plastic polymers has revolutionised casing manufacture.

Most widely used is ABS (acrylonitrile butadiene styrene), which can be carefully extruded under very controlled and monitored conditions to provide industry standard 3.06 m lengths with an OD of 70 mm and an ID of 57.8, and a dimension of 61.25 mm across the groove bases, which are straight and may have a twist (spiral) of no more an $\pm 0.5°$ within the 3 m length. This spiral, albeit small per sample length, could lead to cumulative errors along a deep (say 100 m) borehole. The extrusion process is arranged to minimise this spiral, but it may be necessary to conduct a spiral survey of the completed installation at depths greater than 30 m.

A similar alternative casing with an OD of 58 mm, an ID of 45.8 mm and a dimension over grooves of 45.8 mm is also commonly produced. Other sizes are available throughout the world, most work with a standard measurement unit but at the smaller diameters the curvature limitations commented on earlier apply. At the small OD end of the scale is 48 mm OD casing, and this is not recommended for anything other than structural concrete elements, where little shortening will occur and the reference tube will be fixed to the reinforcement and intimately encased within the concrete matrix before any readings are commenced.

The casing can be used to depths in excess of 30 m and, if lateral deformation occurs, there will obviously be an increase in overall length. In contrast, if major settlement occurs at the same time there could well be a considerable shortening, see Figure 3.43.

As the reference casing is required to match the ground displacements, the casing should be intimately connected to the soil wall by the 'grouting'. Small slippage can occur, but some allowance for telescopic movement at joints must often be incorporated in conditions where serious displacements are anticipated. Arrangements for telescopic connections must also ensure:

a) That the orientation of the reference grooves does not change.
b) Whatever measuring device is employed, it must be able to pass accurately and smoothly from one casing length to the next.

Figure 3.44 shows Soil Instruments' various connecting systems. The standard snap casing (EC Casing) provides ideal groove connections but has little telescoping tolerance. The riveted telescopic joints can provide about ±50 mm change in length, but require additional external waterproof sealing and have a greater diameter over the telescoping length. The telescopic coupling also has wheel grooves in

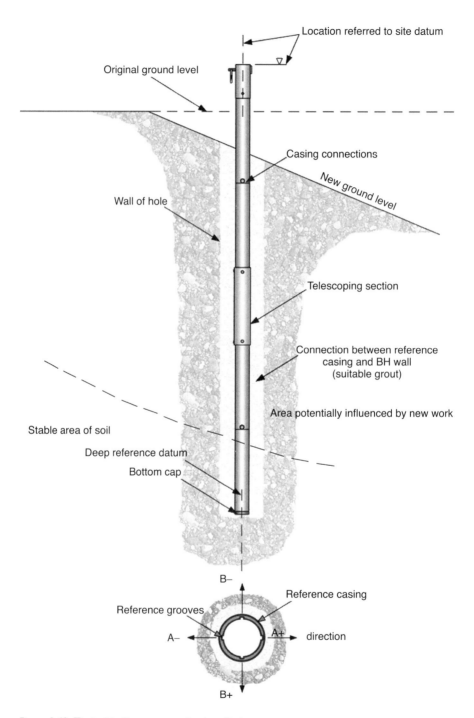

Figure 3.42 Typical inclinometer casing installation.

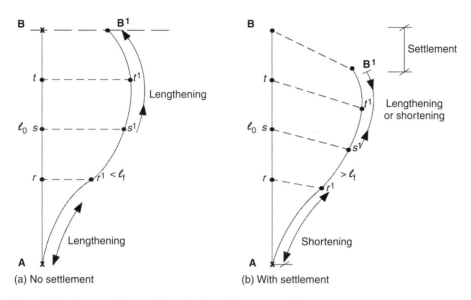

Figure 3.43 Causes of changes in casing length (exaggerated).

the larger diameter, however if one wheel is within a telescoping length the $+A/-A$ measurement will also compensate for this, as shown in Figure 3.44.

The borehole

The sections so far have all described the idealised situation and the factory prefabrication of high-quality components over which the manufacturers have full control. The potential accuracy of these parts of the installation is very high (in the order of 0.05 mm/gauge length or higher). However, the system now needs to be installed in the field, and often in only partially assessed ground conditions and during the early stages of a project when the available site investigation (SI) data are usually limited and site clearance and project organisation is underway.

The quality casing cannot just be 'wished into place' like the nodes in a computer simulation. An open hole has to be formed to full depth (preferably to very stable bed strata) and a completely coupled and sealed case has to be installed. Of necessity, this hole must be larger than the OD of the casing and, depending on the depth, the stratification and the soil types penetrated and the bore size may change with depth. Further construction methods will embrace any of the standard SI rigs, from shell and auger percussion units to full rotary flyte auger and rock drilling units. The resulting hole may be open, cased or partly cased. Water will almost certainly be encountered somewhere within the drill depth and, as with all SI situations, may well be the cause of borehole instability problems. Very commonly the client and his consultants jump on these installation holes as sources of additional SI data. This is actually quite useful, as the knowledge of the actual stratification and soil properties along an instrument line is a considerable help in the later interpretation of deformation data. It does, however, make the process somewhat slower, and this

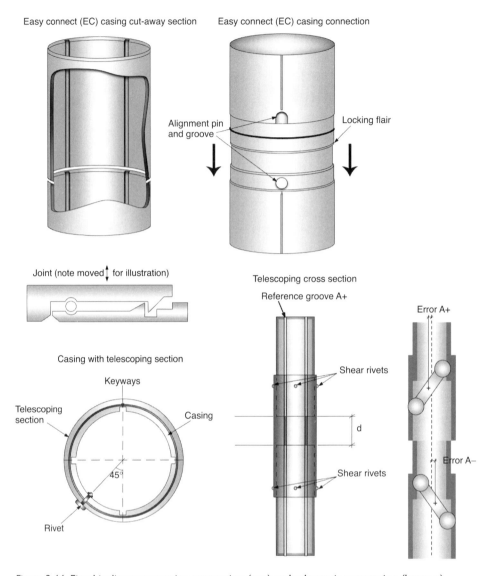

Figure 3.44 Fixed inclinometer casing connection (top) and telescopic connection (bottom).

must be allowed for and any costing discussed with the client prior to commencement of work.

The formation of the hole throughout the full depth of soil and the installation of the casing must ensure that the reference is correctly orientated and acts with the soil. This involves not only the formation of the hole but also casing insertion and grouting. Much of the hole formation will be decided by the soil profile (assessed from prior SI data) and the equipment available. There are different requirements for percussion (shell and auger) rigs with and without casing, and rotary rigs; again, with either simple auger casing or flyte auger hole support, and even, in loose granular materials, with drilling mud support.

The installation

We need to consider the installation stage as it effectively controls the quality of the final output. The insertion of the reference casing and its final connection to the soil matrix must be in a way that *does not* alter the local soil response to the eventual engineering work that is to be monitored. This may sound naïve, but there have been cases where a series of instrument holes have been installed more like mini piles (soil nailing) and gross distortions have occurred either side of the instrument section but not on the section line. While, in other cases, the units have been so poorly installed that local deformations are supposedly only associated with the instruments. The best installation method, and particularly the grouting, will actually be individual to each and every site because the geological conditions are by their very nature unique to each location. It will be necessary to use discretion, placed in the hands of an experienced engineer or technician, who should be allowed a considerable degree of autonomy. The only requirements are that:

a)   Once placed within the hole, the reference tube must be completely surrounded by a continuous matrix of cement/bentonite/polyfluoro acetate (pfa) grout so that there are *no voids* into which a length of casing can later distort.
b)   When set, and the inclinometer system is in use, the grout must move the reference tube within the borehole in the same manner as the surrounding ground.
c)   The set grout must *not* be significantly stronger than the surrounding soil.

(The absolute 'set' strength of this mix is not critically important except in the sense that it should be reasonably representative of the soil shear strength, preferably marginally higher. Grossly higher set strength should be avoided as this will have difficulty in flowing around the tube during pouring a) above, and in its final condition it could have a brittle characteristic and crack locally or even act as a soil nail or pin 'pile').

Complying with these broad requirements can only be achieved by experimentation on site with the particular cements, bentonites, fillers (pfa, etc.) and even water, that are available with the actual geological materials in the specific location as a reference. This is where the experienced site engineer or technician should be allowed to exercise his discretion.

The one requirement that should be specified is that the installer should provide a careful record in a set of notes of exactly what was done both for unsuccessful mixes and for the eventual successful mix, and comments about what end properties he or she sought and what actually occurred in each instrument location, i.e. a record of installation viscosity (visual description, not numeric value); *grout take*: time taken: *depth*, *diameter* and soil profile and when measured; what 3, 7 and 28 day cylinder strength tests have been taken. A volume check should always be made.

This data may be critical during interpretation and for any queries at a later date.

Installation procedure

The open borehole will, as stated, be of various diameters, cased or partially cased, pass through several strata types, and be partially or wholly flooded.

The first step is to install the inclinometer casing. It needs to reach into the 'bed rock' or at least out of the zone of influence, with the +A grooves in the direction of expected movement. It must not suffer any buckle failures and must remain intact during grouting and drill casing removal (if the borehole is lined after drilling). The inclinometer casing comes in approximately 3 m lengths (3.05 m to be precise) which are carefully machined to ensure groove orientation within each length, with the snap-together joints ensuring this is maintained during coupling. The telescoping units (if used) are factory assembled, sealed and with snap joints at each end.

On site, each length of casing is lowered into the borehole and supported while the next length or the telescoping coupling, respectively, is attached. The lowest casing section should have a bottom cap attached to ensure that no foreign debris enters the casing during installation.

This means the casing will be an empty hollow cylinder; hence, once it enters a flooded borehole it will tend to float. This buoyancy can be partially compensated for by filling the inclinometer casing with clean water that can, if desired, eventually be pumped or blown out.

Theoretical concepts behind borehole grouting

It might come as some surprise that there is considerable tolerance in strength of the grouting material. The argument concerning this is as follows.

The inclinometer casing is originally placed within an empty or partially water-filled borehole, it will be 'overweighted' by flooding with clean water (see above) to ensure that it remains fully in contact with the base of the borehole, see Figure 3.45. The casing of any length will flex (mainly at the joints) into a wave type form, touching the sides of the borehole in various places depending on the individual flexibility of the joints and of the casing. This is not of any consequence as the subsequent grouting must fix this alignment and the initial zero (base) readings will determine the actual alignment. All subsequent displacements will be monitored as changes from this initial alignment.

A grout of suitable fluidity will then be injected from the base upward (i.e. tremmied), using the accepted techniques of tremmi concreting (i.e. the supply pipe will be maintained just below the grout surface and slowly raised up the borehole as grouting proceeds). Any drill casing is withdrawn, also slowly, once the grout level is about 1 m above the toe of the casing. This process is shown in Figure 3.45. If other instrument units, such as settlement spider magnets, are installed outside the casing, use of a tremmi pipe may not be possible; in these circumstances the hole may need grouting by pouring from the top with great care.

It should be noted that the density and hence the buoyancy of the grout is greater than water and a small excess downward load must be maintained on the casing.

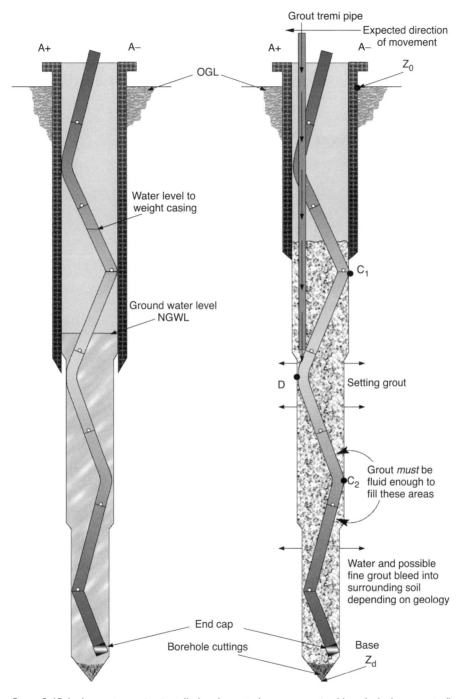

Figure 3.45 Inclinometer casing installed and grouted on an oversized borehole (exaggerated).

Subsequent soil displacement

The soil mass is expected to displace in the $+A$ direction. It is the $-A$ side of the casing that is therefore required to move with the borehole wall Z_0–Z_d. There is direct contact at C_1 and C_2 so the casing will obviously move. The question is what happens between C_1 and D, and D and C_2 (see Figure 3.45)? Looking at the section at D, in a poor installation the casing will be in the situation shown (exaggerated) in Figures 3.46 and 3.47.

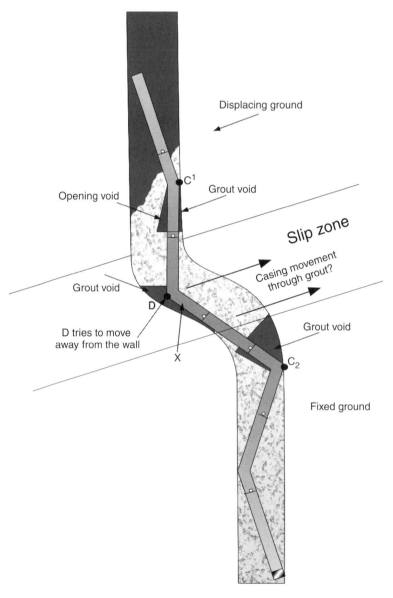

Figure 3.46 Very poorly grouted hole with grout voids shearing on slip zone through D.

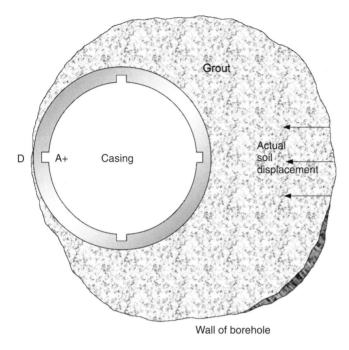

Figure 3.47 Inclinometer casing/grout relationship where the casing is on the side not being pushed.

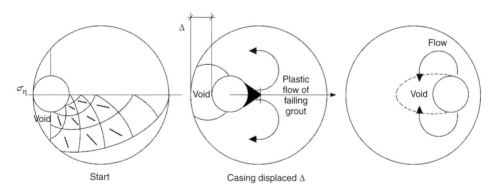

Figure 3.48 Simplified plastic flow of settlement round inclinometer casing (very unlikely).

In a grossly exaggerated diagram, if the casing does not move with the wall because of its stiffness it must be displaced through the grout – idealised as in Figure 3.48.

This is in fact a bearing capacity problem in the lateral direction assuming the grout is a constant strength (S_u) material.

The highest bearing stress value would be $2\pi(S_u + 2S_u)$ or approximately $8.3\ S_u$. If there is a grout void as shown in Figures 3.46 and 3.48, this value can fall to $(2+\pi)S_u = 5.1S_u$ or less (where S_u is the shear strength of the grout.)

The lowest force required to cause this would be generated under the circumstances shown in Figure 3.46. The worst case would be at X with clear opening voids occurring at C_2 and a low bearing value at D.

It can be clearly seen that the critical factor is to ensure there are no void pockets in the vicinity of the wall contact points. Conversely, if the grout is over-strong, like a concrete pile or a soil nail, then the soil itself could fail around the concrete pin. If the tube has 70 mm OD and the grout strength is 500 kN/m^2 – very low (normally it is difficult to produce grout that does not harden to less than 1500 kN/m^2), the lateral force to push the tube through the grout is approximately $(5.1 \times 500 \times 70/1000 \times 1) = 18$ tonnes/m of length. If the soil strength is 200 kN/m^2 (e.g. strong London clay) the force to push a 100 mm diameter grout column through the soil is $(8.3 \times 200 \times 100/1000 \times 1) = 16.6$ tonnes/m of length. But for an inclinometer in soft estuarine clay or peat (shear strength 20 kN/m^2) the value would be only 1.7 tonnes/m length of shaft. It can be seen that over-strong grout can be critical on soft clay sites.

Grouts are rarely as weak as 500 kN/m^2 after 28 days. It is therefore extremely unlikely that the reference tube can ever be pushed through the grout, so it must always remain referenced to its original location within the grout column. It is almost impossible to push the grout column through the soil as the column would shear across its diameter at location X long before the soil became plastic.

It is extremely difficult to generate lateral forces anywhere in the inclinometer casing length sufficient to cause relative translational displacement Δ at D, and it is not necessary to produce stiff, high-strength grout.

Much more important is to provide a grout that during installation is fluid enough to ensure that no voids occur in the critical areas where the casing contacts the wall. When the grout sets it should *not* shrink to be a loose fit within the borehole walls, and on occasion expansion agents have been added. There are no magic formulae for this as each mix design will vary according to the specific supplied materials.

If inclinometers are used for structural monitoring, the casing will normally be wired to the main reinforcement plane. The attachment must be substantial enough to withstand the rigour of live concrete pouring and power vibrators. If the attachments are to be installed within an existing concrete wall, a rock drill hole will be cored at a location avoiding cutting primary reinforcement and will be grouted with a high-strength grout material to match the *in situ* concrete strength.

In-place inclinometers

In special situations where detailed monitoring of high quality and repeatability is required at close time intervals, e.g. three or four inclinometers down to say 35 m depth all requiring reading at six-hourly or more frequent intervals. Then a system of several mini-inclinometers can be installed, usually over 1 or 2 m gauge lengths as a chain – see Figures 3.49 and 3.50.

The in-place inclinometer (IPI) consists of a body bar, which can be varied between 0.5 and 3 m individual gauge lengths, at the end of which is attached a read unit with one fixed wheel and one sprung wheel, and a precision ball-joint coupling at the top which connects into the wheel unit of the sensor above. The sensors can be read instantaneously and provide an individual tilt output for each gauge length. This will provide a complete instantaneous set of data for the whole length of the IPI string.

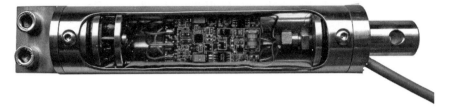

Figure 3.49 Cutaway view of an in-place inclinometer reading unit.

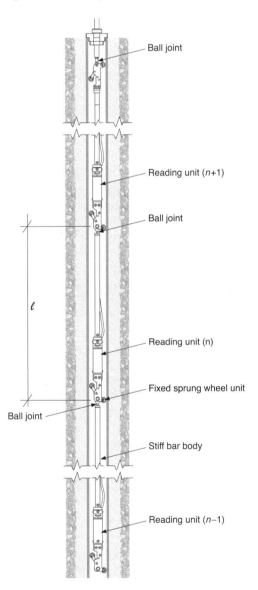

Figure 3.50 Chain of in-place inclinometers from the top of the borehole.

In these circumstances the most important additional data is that rates of movement can be examined, i.e. Δx/hour (or even minute) and a potential failure can be seen to be stabilising or accelerating towards a failure, potentially giving a safety warning, or (if acted on early enough) enabling remedial action to be carried out.

Under circumstances of progressive failure, such a system can continue to work even under gross distortion. The only sacrifice is that the chain of sensors sometimes cannot be recovered under these extreme conditions. The system can also be planned so that casing joints do not influence the reference wheels.

Horizontal profile gauges using an inclinometer probe or IPI units

The monitoring equipment and the use of grooved reference tubing can easily be applied for vertical settlement measurements both within soils and within structural foundations by rotating the system by approx 90°.

The key differences are that:

a) For above original ground level constructions such as embankments and dams and even behind backfilled retaining walls, the reference tube can be placed directly within trenches in the soil mass as construction proceeds, the trenches being carefully backfilled with correctly compacted material. Axial displacement plates (see Figures 3.51 and 3.52) can also be easily installed along the trench for horizontal movements.
b) For below original ground level data, the trench technique can be installed down to a 2 or even 3 m depth below formation, but thereafter instrument holes have to be drilled with directionally controlled drilling equipment near horizontally, usually with continuous casing from an instrument pit or a convenient shaft. The reference casing will lie on the bottom of the drilled hole and has to be inserted so its key (+A and −A) axis is vertical. The grouting is quite difficult and it is often preferable to slope the hole very slightly downward (5°) to prevent a void occurring in the grout along the crown of the instrument hole.

It is very difficult to pull or push an inclinometer along such holes. IPIs which are axially rigidly connected can be installed and later extracted. Figures 3.51 to 3.53 show typical horizontal profile gauge (HPG) installations for horizontal profile measurement, the most important of which is probably the real-time monitoring of compaction or compensation grouting with new tunnel construction in urban or city areas, Figure 3.54.

Interpretation

All inclinometers and IPI systems are excited and the data brought to the accessible end of the casing by screened and waterproof wiring. These data are generated in two forms. The single inclinometer (in the most advanced systems) is digitised within the probe and the digital signal transmitted to a data transmission hub on the cable drum. These digitised data are then transmitted by short distance radio to a personal digital assistant (PDA) or similar device on or near the operator. The operator is responsible

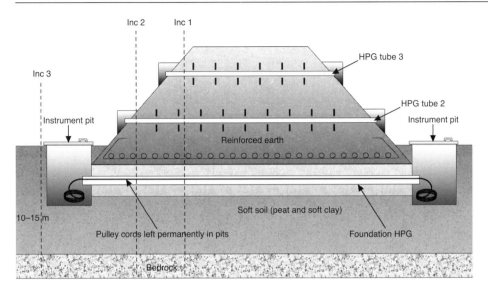

Figure 3.51 HPGs in a road/railway embankment.

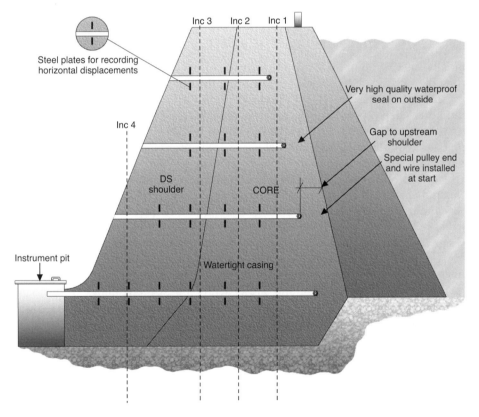

Figure 3.52 HPG in a dam body.

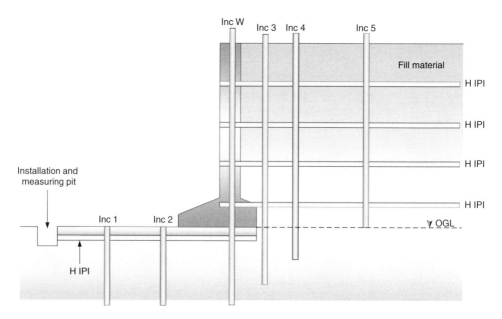

Figure 3.53 Inclinometers and horizontal profile systems in retaining wall construction.

for initially entering all of the basic information, identifying the position and name of the borehole to be read, the date of reading and the time, and the groove axis of the leading wheel (i.e. $+A$, $-A$, $+B$, $-B$). The operator is then responsible for ensuring that each and every depth position is recorded in sequence from bottom to top. Readings at each measurement point are initiated by either a button on the readout PDA, or by pressing a 'key-fob' similar to a car locking key; the user is prompted (if needed) by software in the PDA to move to the next measurement point.

In the case of IPI sensors, the use of an analogue to digital (A/D) converter for each instrument would render the system excessively costly. IPI data (and most older style inclinometers) are therefore analogue, amplified in the reading unit (probe) and transmitted to the reading end of the system as a two- or four-wire analogue signal. Data acquisition is performed by commercially available dataloggers, of which by far the most popular is the robust Campbell Scientific CR800/CR1000. The logger powers and excites the sensors, signal conditions the subsequent output, performs an A/D conversion and subsequently stores the readings with a date and time stamp. The data can be downloaded at convenient intervals by a visiting technician or interrogated by remote means such as radio, cellular network or satellite uplink.

Data reduction

The data are now available as a data file within a logger or PDA; the data consist of:

a)   The instrument reference and the gauge length constant used.

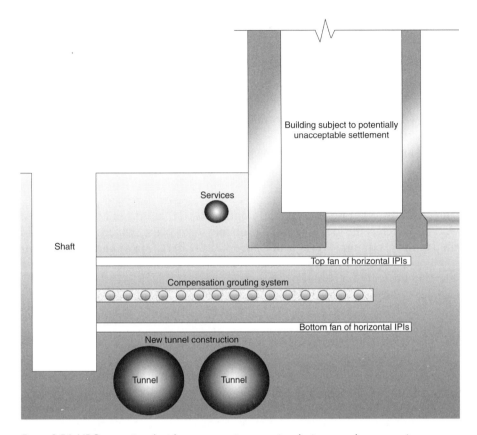

Figure 3.54 HPGs associated with compensation grouting during tunnel construction.

b) The date and time.
c) A set of data pairs consisting of the location within the inclinometer case and the current inclination in terms of $\Delta\theta$ with respect to the gravity vector for each axis (+A and −A).
d) A single $\Delta\theta$ value for horizontal in-place inclinometer chains.

Software loaded onto a computer can now combine two sets of data at one date and time, i.e. +A and −A, and give the mean data pairs for the A-axis, e.g.:

Location 29 m to 28.5 m depth, current inclination +1.5° (rotation clockwise to +A)

The orthogonal mean angle is also normally recorded giving, e.g.:

Location 29 m to 28.5 m depth, current inclination 0.2° anticlockwise to −B

From these data the relative lateral displacements in x and y directions at 28.5 m compared to 29 m can be determined.

The complete mean data sets for one date and time are now available as a block. This will provide the orientation of the casing from bottom to top – *this is a snapshot of the real casing shape at one time* ($t_{current}$) *and is not useful as such*. This block of data needs to be compared to the same block of data at the start or at some key, previous date. *The change from* t_0 *to* $t_{current}$ *is the important data pair*. The earlier orientation is therefore subtracted from the current one, giving data set θ_A from $time_0$ to $time_{current}$ at 0.5 m intervals from the bottom, and as displacement Δx_{0-t} increments to the top of the hole. Two forms of this are important plots:

1) Δx or $\Delta\theta$ against depth, as was shown earlier in Figure 3.29(a).
2) Using $\sum(\Delta\theta_A \times 0.5 \text{ mm})$ the plot of $(\sum)\Delta x_A$ against depth, as was shown in Figure 3.29(b).

The key features to note are: a) the depth location of the maximum $\Delta\theta$ in Figure 3.29(a) which is the location of the probable shear plane (which in turn should correspond to the maximum +A slope in Figure 3.29(b); b) the maximum physical shift of the soil block, and c) the backward slope in the upper sections of the inclinometer's data indicating the solid body block rotation.

If a number of inclinometers or HPGs are available (as in Figures 3.52 and 3.53) – most commonly in slopes, embankments and wall construction – then some additional information can be derived.

Assume a simplified, idealised, circular, slip failure is developing in an embankment system, shown in Figure 3.55.

Simple analysis of the solid block's b–h–e–d–c movement will show that if the system rotates about 0 on a uniform shear zone a–b–h–e–f–a, then the inclinometer system will show contours plotted on the original section as in Figure 3.56. If settlement monitoring is also available this would appear as the contours shown in Figure 3.57.

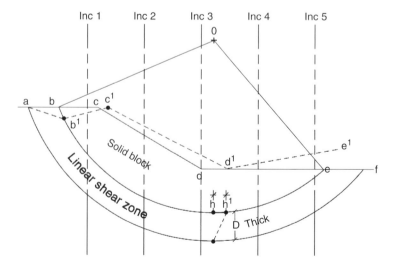

Figure 3.55 Very simplified slip circular failure.

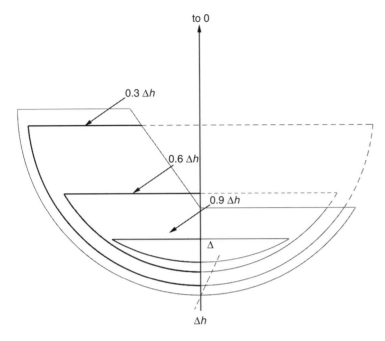

Horizontal movement contours

Figure 3.56 Horizontal movement contours for an idealised circular slip (Figure 3.55).

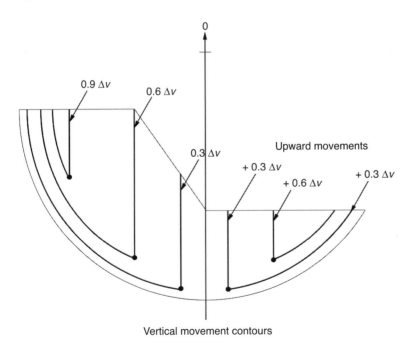

Vertical movement contours

Figure 3.57 Vertical movement contours for an idealised circular slip (Figure 3.55).

Real inclinometer systems incorporated in the embankment situation will unfortunately not provide such ideal contours of the type shown in Figure 3.56. In practice during construction they will be smoothed, as a broad shear band will be mobilised with a design shear failure of 1.4. Shear failures do not occur until late in a failure phase, often starting as fracturing at the top in the location a–b in Figure 3.55, a single rupture working downward from Inc 1 to 2 to 3, with multiple failure planes in the vicinity of 4. The real safety is often provided by the toe area (Inc 4 and 5 area) and digging a toe ditch can precipitate disaster. The real, rather than idealised, data would appear as in Figure 3.58.

However, the lines 0–0 x,y,z can often be determined and the band of slip circles that should be back-analysed would lie within the band x_1 $x_2–y_1y_2–z_1z_2$. The zone above can also be used to identify the area between Inc 3 and Inc 5 on which a berm could be built, and how deep and where to locate shear piles or to build drainage ditches.

In a typical, cantilever, active retaining wall situation, the key mechanism will not be rotational slip but an active frictional mobilisation behind the wall and a passive mobilisation in front of the wall as idealised in Figures 3.59 and 3.60, with a complex active/passive zone below the point of rotation; this can rarely be identified or measured in field situations.

These idealisations suggest the location of inclinometer systems, ideally five or more. If compatible settlement Δz information is available, as suggested in Figure 3.53 then

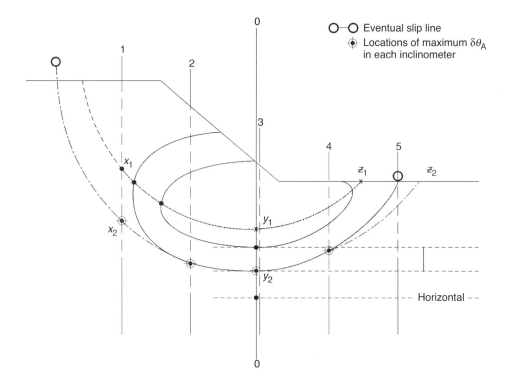

Figure 3.58 Realistic contours of horizontal displacement.

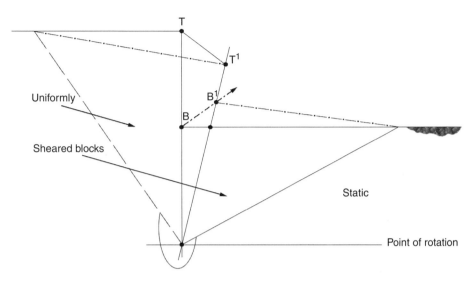

Figure 3.59 Idealised deformed blocks for a cantilever retaining wall.

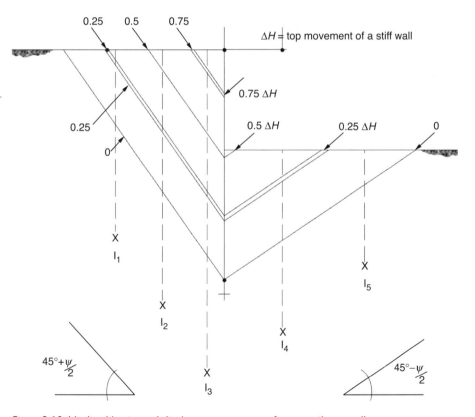

Figure 3.60 Idealised horizontal displacement contours for a cantilever wall.

considerably more analysis can be done giving full details of the areas of maximum shear and the potential mechanisms of failure (see the section on advanced uses below). It should be noted that, even with 'safety factors' of two, the eventual mechanisms can usually be seen to be mobilising at working loads.

The horizontal displacement contour in this case shows uniformly spaced parallel orientation, usually inclined at $45° + \psi/2$ (the dilation characteristic, see later) giving Figure 3.60.

Advanced uses of combined horizontal and vertical displacements

With a fully integrated system of three of more inclinometers and settlement systems (as illustrated in Figure 3.61), the data can be processed in a manner similar to those used in a 'finite element' calculation. Taking the element identified by reference points 4A, 4B, 3B, 3A (see Figure 3.61) the original rectangle 2 m × 4 m has both moved and distorted. The before and after situation is shown in Figure 3.62.

Three actions have taken place:

a) Bodily displacement.
b) Body rotation.
c) Shear and volumetric deformation.

These can best be analysed by using any three of the four available sets of data, i.e. each one of the triangles shown in Figure 3.63.

For each triangle:

i) The coordinates of the centroid of (4A–4B–3A) is found and the centroid of (4a–4b–3a) is also found.

The difference is the bodily displacement of each original triangle $(\Delta x_{(i)}, \Delta z_{(i)})$, Figure 3.63. For the rectangular area therefore: four points with Δx and Δz values are obtained and for the whole three-inclinometer mesh, contours of the displacement intervals (Δx and Δz) can be drawn using a contouring package, see Figure 3.74, later.

Further data for each triangle can also be assessed by taking the vertical and horizontal axes, i.e. 4A to 4B and 4A to 3A (for triangle i) and computing the rotation.

ii) Overall rotation with respect to 4A–4B of the horizontal line, see Figures 3.64 and 3.65

$$\frac{(\Delta Z)_{4B} - (\Delta Z)_{4A}}{\text{length 4A to 4B}} = \frac{\text{relative clockwise rotation settlement of B}^{(a)}}{\cong 4m} = (\alpha \; AB)_{4.4}$$

Overall rotation of the vertical line 3A–4A to 3a–4a

$$\frac{(\Delta X)_{4A} - (\Delta X)_{3A}}{\text{length 4A to 3A}} = \frac{\text{relative horizontal clockwise movement of 4 with respect to 3}}{\cong 2m}$$

$$= (\alpha AA)_{3.4}$$

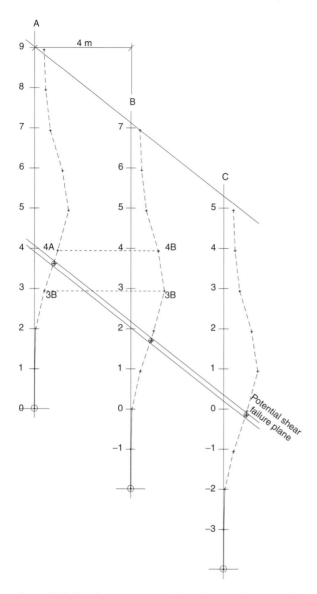

Figure 3.61 Displacement points from three inclinometer/settlement systems.

The bodily rotation of the triangle is the mean of these values, i.e. $\frac{(\alpha\,AB)_{44}+(\alpha AA)_{4.3}}{2}$ $= \alpha_{body}$

The Mohr circle of strain

Before continuing with this ideal triangular element, it is necessary to introduce the 'Mohr circle of strain'.

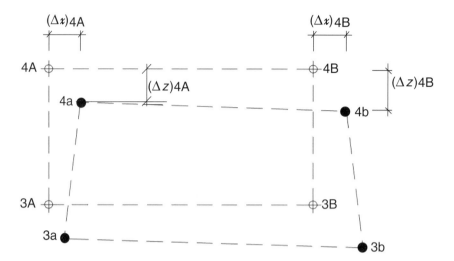

Figure 3.62 Displaced rectangular element from Figure 3.61.

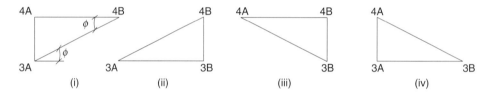

Figure 3.63 Four original triangular elements.

The Mohr circle of stress was introduced in Chapter 2, Figure 2.7, and repeated in Figure 3.68 where the soil mechanics sign convention: compression stress positive, anticlockwise shear stress positive was introduced. There is a corresponding Mohr circle of strain which, when handling inclinometer and settlement data, can prove a very powerful interpretation tool. Figure 3.66 repeats the picture of an infinitesimally small, square element in equilibrium, orientated θ to the horizontal. Figure 3.68 shows the corresponding Mohr circle of stress with θ and $\theta + 90$ data marked on the circle. The 'origin of planes' is found using a plane on which a pair of stresses are known $\sigma'_{\theta+90}$, $\tau_{\theta+90}$ point p drawing this plane directly onto the Mohr circle through the stress value (p) to the opposite side of the circle at P. Hence the principal stress plane $\sigma'_1 - P$ and the two maximum friction (obliquity) planes, (R–P and S–P) are obtained.

Figure 3.66 also shows the outline of the distorted element. The normal displacements and shear distortions are marked as well. In the $\theta + 90$ direction, the normal displacements are compressive p–p' and r–r' and in the θ direction they are extensions q–q' and s–s'. The shear stresses cause angular distortions on the $\theta + 90$ axis, the line r–p rotates clockwise to r″–p″, while on the θ axis the line s–q rotates anticlockwise to s″–q″. The change in the 90° angle (p–q) to p″–q″ is the engineering shear strain

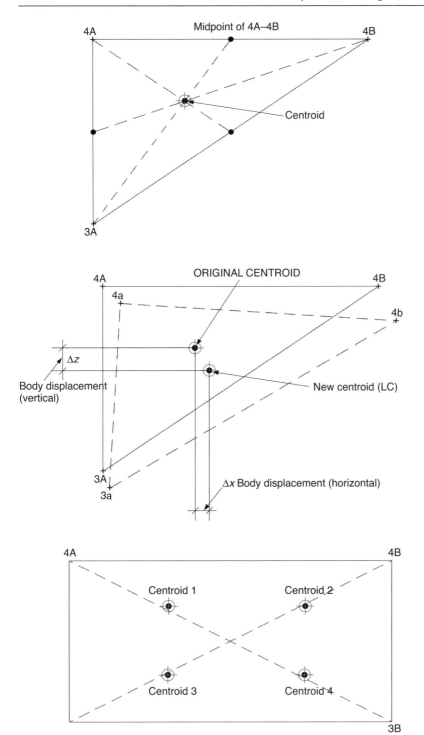

Figure 3.64 Bodily displacement of the triangular element.

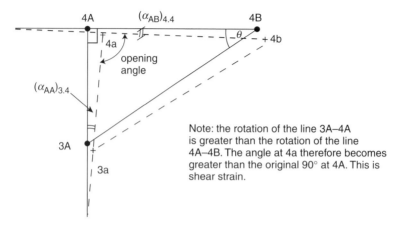

Figure 3.65 Bodily rotation and shear distortion of a triangular element.

Note: the rotation of the line 3A–4A
is greater than the rotation of the line
4A–4B. The angle at 4a therefore becomes
greater than the original 90° at 4A. This is
shear strain.

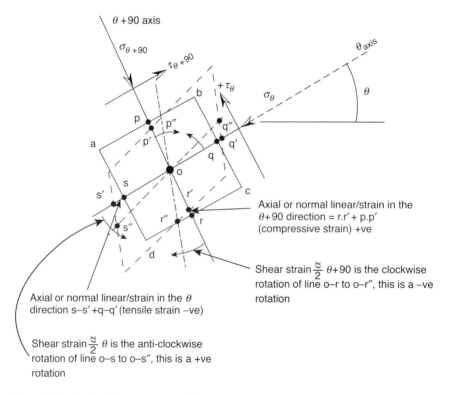

Figure 3.66 Idealised shear stress and shear strains in a square element.

Axial or normal linear/strain in the
$\theta+90$ direction $= r.r' + p.p'$
(compressive strain) +ve

Shear strain $\frac{\alpha}{2}\ \theta+90$ is the clockwise
rotation of line o–r to o–r″, this is a –ve
rotation

Axial or normal linear/strain in the θ
direction s–s′ +q–q′ (tensile strain –ve)

Shear strain $\frac{\alpha}{2}\ \theta$ is the anti-clockwise
rotation of line o–s to o–s″, this is a +ve
rotation

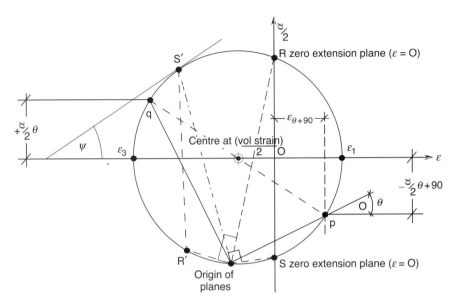

Figure 3.67 Mohr circle of strain.

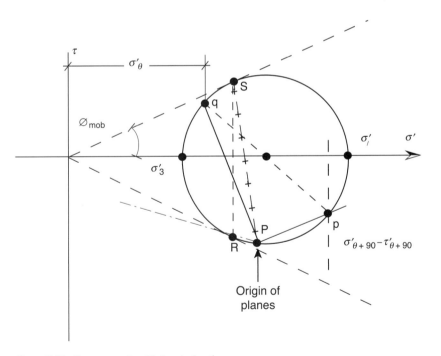

Figure 3.68 Corresponding Mohr circle of stress.

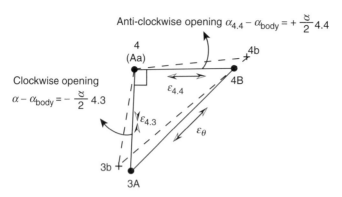

Figure 3.69 Nominal strain and shear within the triangle 3A–4A–4B when all body displacements are removed.

written $\eth$, and hence a component $\frac{\eth}{2}$ is relevant to each plane. Anticlockwise rotation is again termed positive (e.g. point q–q'' rotates anticlockwise, hence the shear $\frac{\eth}{2}$ is positive for plane c–b). The resulting Mohr circle of strain is shown in Figure 3.67.

Returning to the idealised triangular element, from the inclinometer/settlement data discussed earlier, if the bodily rotation components are removed and the point 4a is placed on top of 4A, the deformed triangle will appear as in Figure 3.69. It can be seen that $\varepsilon_{4.3}$ is compressive and $\alpha_{4.3} - \alpha_{body}$ is a clockwise rotation (negative) while $\eth_{4.4}$ is tensile and $\alpha_{4.3} - \alpha_{body}$ is an anticlockwise shear rotation (positive). These two components are the relevant $\pm\frac{\eth}{2}$ components. These data will enable the Mohr circle of strain shown in Figure 3.70 to be constructed.

Again, the use of planes is a critical concept for obtaining important data, to obtain the 'origin of planes' the horizontal plane *across* which $\varepsilon_{4.3}$ acts is drawn from the point X and the vertical plane *across* which $\eth_{4.4}$ acts is drawn from Y These intersect the circle at P (the origin of planes).

iii) The difference between these two angles represents the change in the original 90° angle and hence represents the engineering shear strain (γ_{xz} and γ_{zx}) experienced on the horizontal and vertical planes, see Figure 3.69.

iv) The linear strain on each side of the triangle can also be calculated from the change in length of the lines $\left(\frac{\ell_{4A\ to\ 4B} - \ell_{4a\ to\ 4b}}{\ell_{4A\ to\ 4B}}\right) = \varepsilon_{(4A/4B)} \cong \varepsilon_x$ etc. $\varepsilon_{4A/3A} \cong \varepsilon_y$ and $\varepsilon_{3A/4B} = \varepsilon\theta$.

The value of $\tan\theta$ being obtained by trigonometry $\tan\theta = \frac{\text{length 4A to 3A}}{\text{length 4A to 4B}}$, i.e $\cong \frac{1}{2}$[1] for the original triangle 4A–4B–3A.

There is now sufficient information to construct a Mohr circle of strain (Figure 3.70).

The principal plane across which ε_1 acts can be determined by joining P to the ε_1 point at E. The direction normal to this plane represents the key compressive direction.

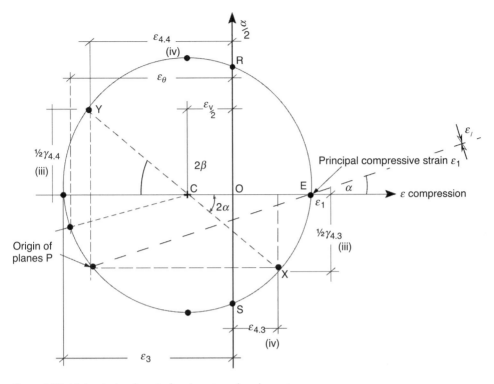

Figure 3.70 Mohr circle of strain for the triangular elements.

Thus three pieces of vital information can be obtained from the data for one triangle:

(1) The location of the circle centre at C. The distance O to C = ½ volume strain ε_v.
(2) The diameter of the circle = the maximum shear strain δ_{max}.
(3) The value of the angle 2 E–X = 2α, the angle α gives the plane across which the principal compressive strain acts with respect to the horizontal plane.

Two other planes are important: the 'planes of zero extension' (i.e. planes that experience *no normal strains*, only shear strain); the points where $\varepsilon = 0$ have been marked as points S and R in Figures 3.70 and 3.71. The two planes *across* which no normal strains occur are the planes P–S and P–R (see Figure 3.71). The normals to these planes P–S' and P–R' are 'directions of zero extensions', i.e. lines *along* which there are no changes in length, only shear slip. Sheild's criterion claims that these two directions are those along which failure ruptures must eventually occur.

The author believes this dilation criterion is critical as it fits both the experience of fully drained failure (where the planes will be at 90 + ψ and fully undrained failure (where the planes must be at 90°). Whereas the concept of planes of maximum

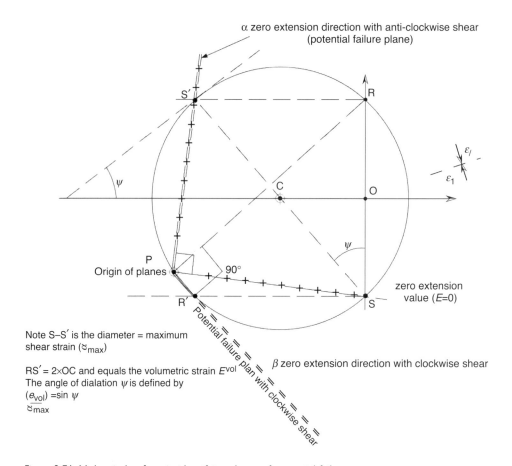

Figure 3.71 Mohr circle of strain identifying planes of potential failure.

obliquity breaks down in the undrained case (ruptures being at 45°). Friction-controlled failure planes fail to agree with Terzaghi's effective stress concepts which, as pore water pressure can have no influence on shear stresses, still requires that failure planes should be at $45 + \frac{\phi}{2}$ to the σ'_1 plane.

These two directions (Figure 3.72) are the key evidence from displacement instrumentation data. They can indicate the areas of critical behaviour and orientation of probable failure even during early construction.

d) The orientation of the two zero extension directions α and β suggest the potential slip planes.

The key information is best presented as:

(i) The basic displacement data δx and δz contours suggested earlier.
(ii) Contours of shear strain magnitude γ_{max}.
(iii) Two zero extension directions α and β drawn as vectors.

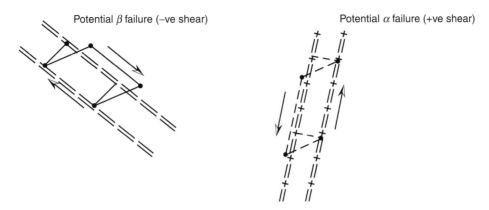

Figure 3.72 Shear deformation on failure planes.

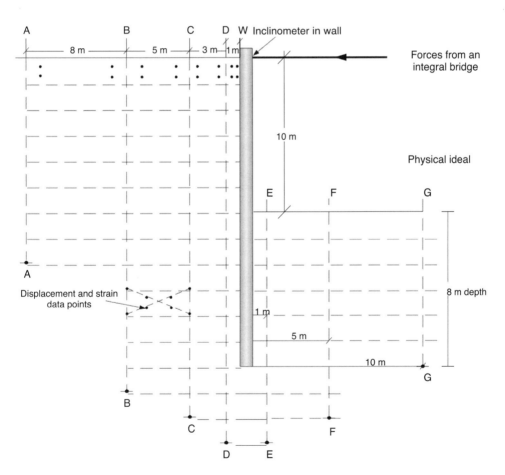

Figure 3.73 Elements available from a comprehensive array of inclinometer/settlement units (7 in number).

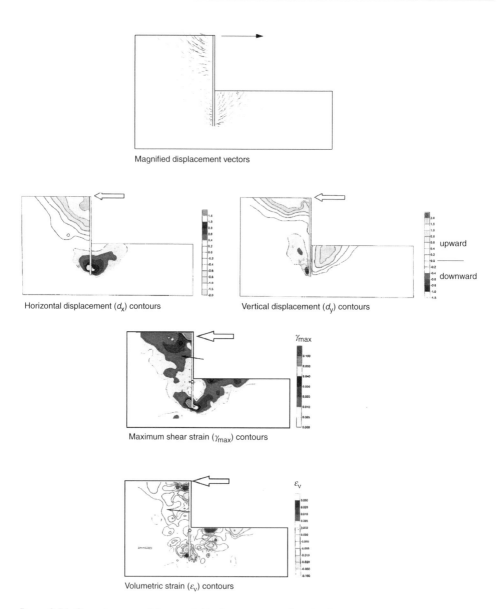

Magnified displacement vectors

Horizontal displacement (d_x) contours

Vertical displacement (d_y) contours

upward

downward

Maximum shear strain (γ_{max}) contours

γ_{max}

Volumetric strain (ε_v) contours

ε_v

Figure 3.74 Complete set of data available from a comprehensive instrument system.

Full output information from combined inclinometer/settlement systems is best visualised by physical plots, which are shown for a classic soil/structure situation:

(a) Passive retaining wall toe cantilever scheme (Figure 3.73).
(b) Data presentation in sections (Figures 3.74 and 3.75).

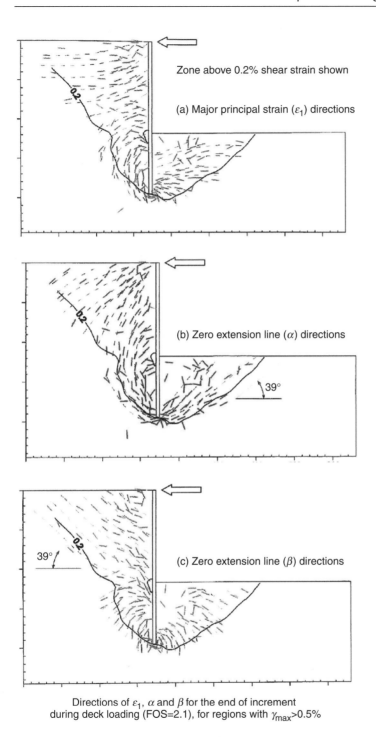

Zone above 0.2% shear strain shown

(a) Major principal strain (ε_1) directions

(b) Zero extension line (α) directions

(c) Zero extension line (β) directions

Directions of ε_1, α and β for the end of increment
during deck loading (FOS=2.1), for regions with γ_{max}>0.5%

Figure 3.75 Complete set of data available from a comprehensive instrument system.

Chapter 4

Vibrating wire instruments and localised measurement of strains

In the previous chapters, physical displacements over extended areas have been determined by mechanical means based on gravity and continuous vertical and horizontal reference lines. Displacements over small gauge lengths directly providing strains can be obtained by the use of vibration frequencies, which can be changed by altering either the length or stress in a tuned wire. This concept has provided a wide range of field instruments which will be covered in this chapter.

Basic mechanics of vibrating wire instruments and linear strain measurement

The basic principle of a vibrating wire gauge is simple and can be compared to all stringed musical instruments. A wire (e.g. a guitar string) responds when excited (in this case by the guitar being played) at a specific fundamental frequency depending on three factors:

a) The length of the wire (string) between fixed points; the longer this length, the lower the frequency.
b) The elastic characteristics and the mass of the wire/unit length of wire; the heavier the wire the lower the frequency.
c) The tensile force in the wire; the higher the force, the higher the frequency.

If a measuring unit is to be constructed using a vibrating wire system the length of the wire will normally be physically fixed, as will the mass and the elastic characteristic. The instrument variable will therefore be to change the tension force in the wire. Guitar

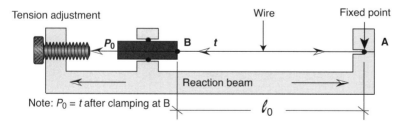

Note: $P_0 = t$ after clamping at B

Figure 4.1 Components of a vibrating wire unit.

strings are all almost identical in length; each one has a different mass/unit length but each is tuned to a specific fundamental frequency by adjusting the tension. All vibrating wire instrument systems comply with this concept. This arrangement will be discussed first. The alternative is to fix the initial tension and then use very small length changes, which due to the area and elastic properties of the wire, will change the tension and hence the frequency.

The key components are shown in Figure 4.1.

The key components are:

a)   The wire length l_0.
b)   The reaction beam or frame.
c)   The application of tension P between the wire and the reaction beam and varying this by some external action.
d)   An initial tension t to set a zero is usually applied at the fixed end A. If the frequency of vibration can be measured, then the value of t can be determined. The wire is normally a high-quality steel wire. A basic principle of electromagnetism states that if a metal wire moves in the near vicinity of a coil of wire, an electromotive current is induced in the coil. Contrarily, if a current is in turn passed through the coil, then the wire will be displaced by the same electromotive forces.

A small induction coil placed close to the centre of the wire can therefore be used to do three things:

a)   A short 'burst' of AC current near to the natural resonant frequency of the wire will stimulate the wire into vibrating at its natural fundamental frequency.[1]
b)   When this excitation is switched off, the wire will settle down to its natural frequency which will steadily decay in amplitude depending on the damping characteristics of the system.
c)   While the wire is vibrating, the coil will respond with an identical AC frequency to that of the fundamental frequency of the wire (due to the electromotive response). With a high accuracy electronic time base, a number of cycles (normally about 200) can be counted against time and hence the period determined (defined as the period in time a single 'sweep' of the fundamental frequency waveform takes).

If the system is computer controlled, this excitation can be carried out a number of times, the mean value will be a more accurate answer. In practice, the response is not calculated on a theoretical basis, but calibrated directly in a bench test giving a P against frequency or period calibration chart for the practical set up (note: due to the extreme repeatability of modern manufacturing processes and electronics this is often expressed as a 'batch' factor across many tens or hundreds of single gauge units of the same design).

Real design considerations

The key elements are:

a)   The length of the wire.
b)   The structure of the reaction beam and the fixed ends.

c)   The placement of the excitation/reading coil.
d)   How the initial t is set during the initial construction (manufacturing) of the gauge.
e)   How t changes during use due to the external displacements or load changes that are to be measured.

These requirements determine the whole detailed design, the subsequent calibration and eventual use.

The elements listed above form the basis of all vibrating wire units and conventionally take the form of:

a)   The wire: a high-quality steel wire (not unlike a piano wire) the length of which is determined by two swaged or clamped collars.
b)   The reaction beam: a stiff tube of stainless steel surrounding the wire and providing clamping, and the initial t value at the fixed end A in Figure 4.1.
c)   The tension in the wire which is initially set (by a number of methods) and is subsequently varied by the external means.

The simplest method of altering the tension in the wire when changes in pressure are to be measured is to incorporate a deformable diaphragm, clamped at the end B in Figure 4.2.

The whole unit is shown in Figure 4.3 and a diagrammatic section through the diaphragm displacements with changes in pressure is shown in Figure 4.4.

For the diaphragm layout (see Figures 4.2–4.4). The diaphragm is a carefully machined unit; its diameter and thickness determine the load range of the instrument (in kPa) and match the size and load capacity of the wire. The boss at the centre of the

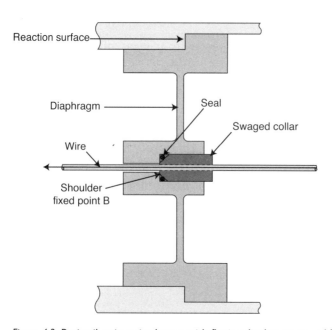

Figure 4.2 Basic vibrating wire layout with flexing diaphragm at end B.

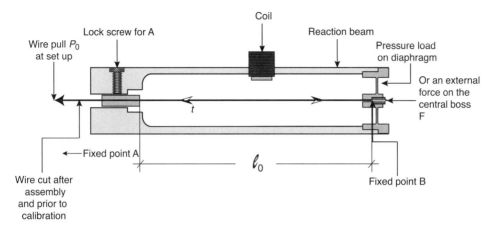

Figure 4.3 Simplified section through a diaphragm unit measuring load or pressure changes at end B.

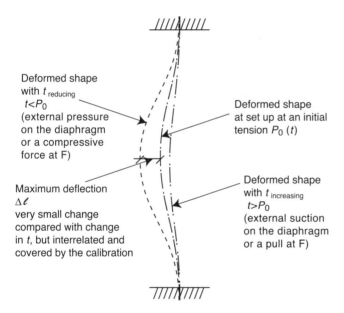

Figure 4.4 Deflection of diaphragm. End B.

diaphragm is clamped to the swaged collar on the wire to form the fixed point B which is also sealed, the wire is threaded into the reaction tube, and the diaphragm and wire are pulled together from end A, by the initial load $P_0 + \max \Delta P$ on the assembly bench, when the diaphragm deflection will be $P_0 + P_{max}$. Finally the lock screw clamps the collet at end A, to the reaction frame with an initial P_0 is left in the wire.

Excess wire is cut away, the diaphragm end is then loaded by applying external pressure to the diaphragm or by a force to the central boss, and the frequency changes are measured as the wire tension falls as the load (pressure) on the diaphragm rises.

The alternative is to physically alter the tension in the wire by applying this tension through a spring system either in tension or compression. Using a spring means that a large displacement can be applied at the remote end of the spring to provide the change in wire tension t. This arrangement amplifies the displacement range, for use of a vibrating wire as a strain gauge; springs of various stiffness are incorporated in the system as shown diagrammatically in Figures 4.5 and 4.6.

Looking at the loading on the diaphragm in detail, the conditions in Figure 4.7 commonly apply when the unit is used to measure pressure.

If either the atmospheric pressure or equal hydraulic pressures initially apply both internally and externally, then only the change in the externally applied force ΔF will

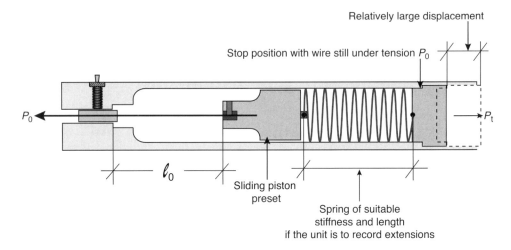

Figure 4.5 Tensile large strain gauge.

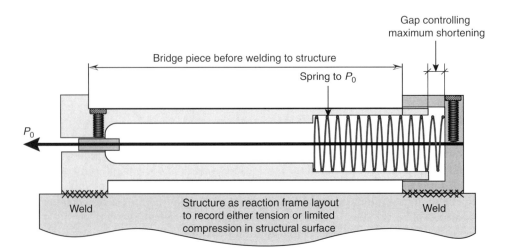

Figure 4.6 Compressive/tensile small strain gauge.

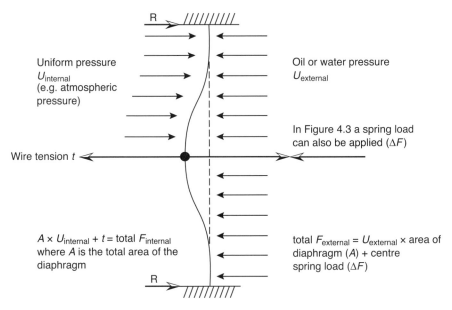

Uniform pressure
$U_{internal}$
(e.g. atmospheric
pressure)

Oil or water pressure
$U_{external}$

In Figure 4.3 a spring load
can also be applied (ΔF)

Wire tension t

$A \times U_{internal} + t =$ total $F_{internal}$
where A is the total area of the
diaphragm

total $F_{external} = U_{external} \times$ area of
diaphragm (A) + centre
spring load (ΔF)

Figure 4.7 Diaphragm under internal and external (applied) pressure.

cause changes in t; this is the normal form of vibrating wire unit used to measure forces.

If the interior of the reaction tube is sealed and evacuated so that $U_{internal} = 0$ (absolute), then the load changes in the wire are directly related to $AU_{external}$ where $U_{external} =$ change in atmospheric pressure + change in water (piezometer) or oil (pressure cell) pressure. (It must always be remembered that atmospheric pressure can change by ± 50 mbar.)

This format is used for the measurement of pore water pressure (pwp) in piezometers or oil pressure (pressure cells) Δu. The internal area when evacuated must never have direct access to atmospheric pressure.

For the displacement spring layout (Figures 4.5 and 4.6) the wire is tensioned, together with a preset load in the spring against a reaction or preset stop(s), and locked to leave a pre-stress P_0 in the system. As the remote end of the spring is pulled off this stop, the spring extends and the wire tension rises, the spring applies a $\Delta P_{external}$ direct to a piston block and through to the wire. The movement of the block will be minute; ideally there must be minimum friction loss in the piston; this is acceptable if the loading is monotonic, if there is load reversal some hysterysis may be experienced. The load–extension relationship is effectively controlled by the load–extension range of the spring. The commonest arrangement is the compressive spring (Figure 4.6), as this allows both increase and decrease in load and displacement.

The complex mechanical details of real vibrating wire units are illustrated in Figures 4.8 and 4.9. They arise due to the vacuum requirement, and also due to manufacturing, sealing, electrical security and calibration complexities.

Figure 4.9 shows a typical piezometer (pwp) unit, the position of the diaphragm is shown and the reaction tube is formed of two components: the cut-away tube and the

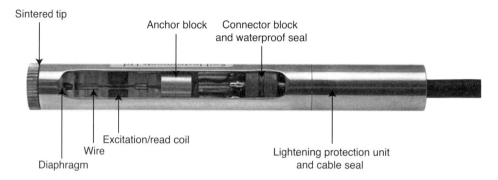

Figure 4.8 Section through a piezometer.

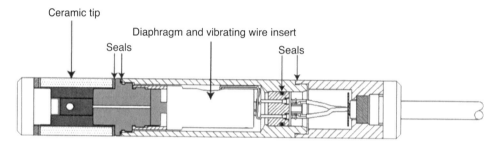

Figure 4.9 Heavy-duty piezometer incorporating a vibrating wire insert.

anchor block. The vacuum is contained within the outer shell and O rings within a wire sealing block. The pore water contacts the diaphragm through the sintered tip (as shown) or a suitable ceramic tip forming an end cover.

The items at the right-hand end all form a waterproof seal for the electrical connection to the readout unit.

The inclusion of a piston and spring component (see Figure 4.10) enables any choice of displacement range to be covered, as the spring constant will enable the small deflections that occur in the vibrating wire to be amplified by the spring to provide a displacement over any change in length up to a displacement of approximately 300 mm.

Load cells and embedded strain gauges using the reaction beam as the strain control

Interesting variations of the vibrating wire system are used for large load cells and embedded concrete strain gauges. These are shown diagrammatically in Figures 4.11–4.13.

In the strain gauge units (Figure 4.11) the whole external reaction frame is formed by a thin-walled tube (usually stainless steel). The wire is stressed within this tube so the unit is self-stressed to a P_0 value. The unit is loosely wired to the secondary

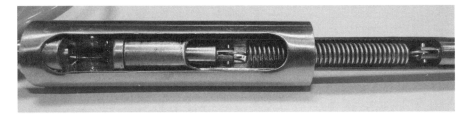

Figure 4.10 A spring-controlled vibrating wire instrument.

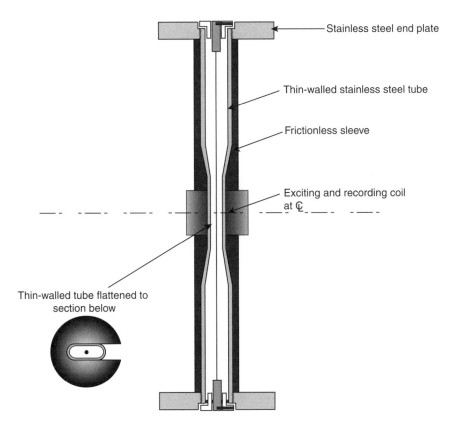

Figure 4.11 Embedded concrete strain gauge.

reinforcement so as to be aligned with the primary reinforcement. There are large-diameter stiff discs welded at either end, the purpose of which is to pick up a small component of the load being transmitted down the pile, resulting in a shortening of the instrument which is the same as the concrete, resulting in a consequent fall in the wire's fundamental frequency. The discs are sleeved between end plates.

With 'sister bars', an actual length of reinforcement is used as the reaction frame (Figure 4.13) from B to C, and the two bond lengths A–B and C–D take stress from the

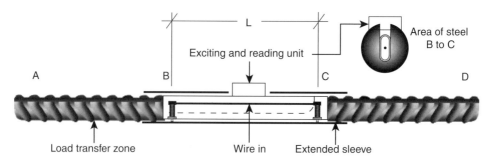

Figure 4.12 'Sister bar'.

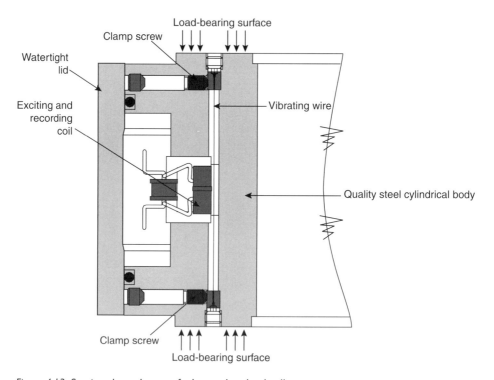

Figure 4.13 Section through part of a heavy-duty load cell.

main reinforcement, working on the assumption that between B and C the instrument will experience the same strain as the whole column. The length B–C is again sleeved to isolate the gauge length from the concrete. Theoretically both Figures 4.11 and 4.12 will modify the load distribution in a section; however, the cross-sectional area of the thin stainless steel tube in Figure 4.11 is so small that it will have a negligible influence. This is not so in the sister bar unit shown in Figure 4.12.

Steel load carried is $\stackrel{\triangle}{=} 15 \times$ same area of concrete. $W_T/(A_s \times 15 + \text{area concrete}) =$ average stress in concrete.

In a normal situation, if gauge length is L the strain (ε_{conc}) in the concrete will be:

$$E_{conc} \times \frac{W_T}{15 A_s + A_{conc}} = \varepsilon_{conc}$$

where E_{conc} is the elastic modulus of concrete.

The change in length Δ which should alter the frequency over the gauge length L will be:

$$\Delta = \varepsilon_{conc} \times L$$

If three sister bars are incorporated in the section, then the effective area of steel increases and reduces the equivalent average concrete stress to:

$$W_T / \{(A_s + 3 \times A_{sister\ bar}) \times 15 + area\ concrete\} = modified\ concrete\ stress$$

And consequently Δ will be modified. A correction may be necessary, which includes the area of the sister bars, to calculate the actual load in the column.

Load and stresses

Figure 4.13 shows a part cross-section of a massive machined steel casting capable of transmitting up to 600 tonnes through the thick circular annulus between the top and bottom bearing surfaces. Three or six small holes drilled through the annulus at 120° or 60° in plan can have vibrating wire gauges installed within them, each with a coil unit at its midpoint. The load cell acts as the reaction frame which is shortened by the applied load, lowering the frequency of the wire as the tension falls. If the wires are read separately, individual changes from the zero load settings can be used to indicate the eccentricity of the load.

Finally, Figure 4.14 shows a thin-walled hydraulic load cell in which two flexible discs of stainless steel are edge welded to form a very thin 'flat jack'. A hydraulic pipe connection is incorporated in the edge and the unit filled with hydraulic oil. A vibrating wire piezometer is attached to this pipe connection and, as it is an enclosed hydraulic system and oil or water is incompressible, the uniform load on the surface of the disc will be transmitted directly to the piezometer unit.

Analysis of stresses to provide full stress analysis

In essence, all vibrating wire instruments are load cells in which change in load carried through a diaphragm or a spring system and hence into the reaction frame, or the load shared between the reaction frame itself and the wire, results in a change in the tensile load within a vibrating wire. The change in tensile load in the wire provides a change in frequency which is measured, and whatever output parameter is required is obtained from a direct (batch) calibration.

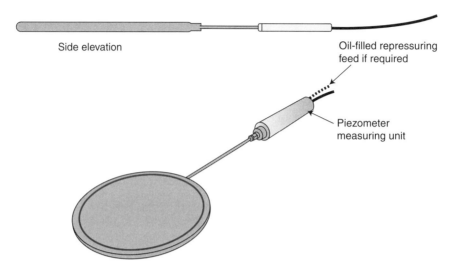

Figure 4.14 Diaphragm (oil-filled) pressure cell.

The interpretation of vibrating wire instruments

Vibrating wire instruments divide into two distinct sets:

a)   Those responding to fluid load changes on the outside of a diaphragm which are subdivided into two groups. The commonest record natural groundwater pressures and guaranteeing connection to ground water can be very difficult. Less common but much more reliable are oil-filled units, factory connected to pressure cells.
b)   Those responding, often via a spring system, to external displacement changes.

Assessment and interpretation of vibrating wire data

The original and still the greatest use of vibrating wire units is for measuring pore water pressure within natural soil or rock strata, or within made ground or fill.

When used to measure fluid pressure against a diaphragm the situation is unique, because fluid pressure in a small volume of soil can be assumed to be uniform and (below the water table) usually continuous. The data represent the actual pressure at the ceramic tip or diaphragm. This is extremely reliable (provided the water is de-aired), but care must be taken to ensure the measurement is either against the ambient atmospheric pressure or against an absolute sealed-in vacuum, which is maintained for the life of the instrument.

The key to obtaining quality data is to ensure that:

a)   The water within the unit between the ceramic tip and the diaphragm is continuous and contains no dissolved air.
b)   The saturated unit and ceramic tip are installed in the soil in such a way that the natural ground water makes intimate contact with the de-aired water within the unit.

It must be remembered, however, that almost all natural ground water systems contain dissolved air to some extent and inevitably some air can migrate into the piezometer system over time, high air entry filter tips can minimise this.

The method of saturating and de-airing the unit is to boil the tip below water, or alternatively, to subject the whole unit to a high level of vacuum, insert the tip into de-aired water and slowly remove the vacuum. The unit will suck in the de-aired water – this operation should be carried out at manufacture. The saturated piezometer should be transported to site in a container (plastic bag) of de-aired water and installed in a water-filled location in the soil with the minimum exposure to air.

Depending on the shape of the ceramic tip and the type of soil into which the unit is to be installed, a number of specialist tools are available for forming a suitable pocket. Every effort must be made to ensure that the soil or the pocket into which the unit is to be pushed or placed is submerged in natural ground water, the unit is lowered down until it and its container are below the natural ground water. The unit should then be pushed out of its container and placed in the pocket or pushed into the undistributed natural soil. Intimate contact cannot always be guaranteed despite every care being taken.

After installation, the cables and connections above the unit (possibly placed within a sand pocket), are sealed with a bentonite plug to isolate the piezometer location from water from a different horizon within the borehole. The borehole may be grouted or plugged to the next piezometer level. If the unit is a simple stand-pipe piezometer for water level monitoring, the remainder of the hole is open or sand-filled. It commonly takes some time, particularly in clay soils, for the natural pore pressure to stabilise, as the insertion of the piezometer will generate some local pore pressures (which may even be negative). Careful observation will be necessary to determine the *in situ* pressure and response time.

Strains

It must always be remembered that the data provided in the strain gauge format is relevant only in the axial direction in which the instrument is set, and will represent a change which is influenced by the external changes modified by the influence of the instrument itself. The gauge lengths tend to be short and local, typically 3 cm to 250 cm (in contrast with in-place inclinometers, electrolevel beams, Bassett convergence systems, etc., where the gauge lengths are 1–3 m). Further, if incorporated in components of complex soil/structure interaction problems, such as large pile groups with heavy pile caps or stiff floor slabs supporting large frame buildings or integral bridges, then the behaviour of points within individual members of such complex structures may be extremely difficult to interpret and often bear little relationship to, for instance, the response of the same instrument when installed in an independent test pile.

The units recording displacements are all in effect linear strain gauges, the two ends of which determine the effective gauge length ($\iota_{effective}$), which is *not* necessarily the length of the wire, and must be directly attached to the material or surface on which the change is to be measured. There must be *no* interference between the two ends. Secondly, it must be clearly understood that what will be measured is the displacement *change* that occurs *after* installation and is:

a)   In one specific direction.

b) Over a relatively short gauge length.
c) Due only to any stress change and consequent strain that takes place *after* the
    strain gauge is fixed in place.

It is very rare to be able to measure strains from a completely unloaded state. Stresses
and strains on most structural items are at least two-dimensional and, even for this case,
three normal strains at known orientations are needed to determine the full situation,
in the manner of a standard strain gauge rosette, Figure 4.15. For stress measurement
in two dimensions, the layout in Figure 4.16 is commonly used.

In a three-dimensional soil mass there would ideally be six linear strains between
four nodes, as in Figures 4.17 and 4.18 for three-dimensional stresses in soil.

The majority of soil problems can be approximated to plane strain (i.e. long
walls, embankments, etc.) and considered as two-dimensional in the key deformation
plane.

It must always be remembered that installing the unit will alter the behaviour in the
vicinity of the unit. The magnitude of any alteration can be extremely difficult to assess.
The same concept applies to the vibrating wire diaphragm pressure cells in Figure 4.14
which only measure normal stress; so again to obtain a proper understanding, three
must be installed: a) horizontal, b) at 30° and c) at 120° to determine the full stress field
(see Figure 4.16); or in a pit as in Figure 4.18 for determination of a three-dimensional
stress field.

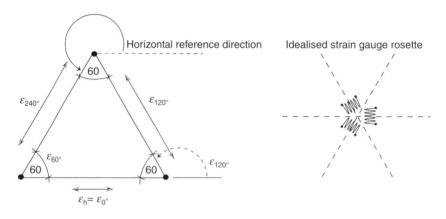

Figure 4.15 Ideal linear strain measurements to determine the full two-dimensional field.

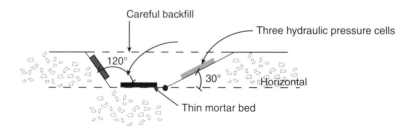

Figure 4.16 Layout of vibrating wire pressure cells in a pit.

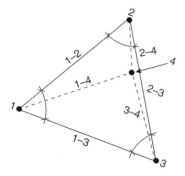

Figure 4.17 Three-dimensional array.

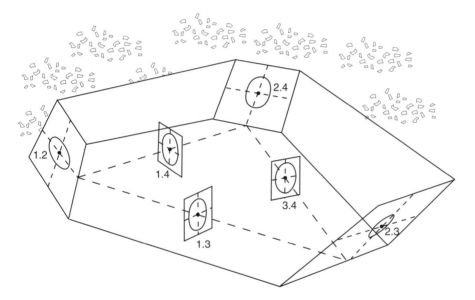

Layout: Three vertical normal pressure cells
Three inclined normal pressure cells

Figure 4.18 Three-dimensional stress measurement in soil.

Data interpretation

The Mohr circle of stress, discussed earlier (Figure 2.7), is repeated in Figure 4.19 and the Mohr circle of strain is shown in Figure 4.20. In terms of stress, three key parameters define the circle:

a)   The centre, mean normalised stress $s' = \frac{\sigma'_1 + \sigma'_3}{2}$.
b)   The radius maximum shear strain $t = \frac{(\sigma_1 - \sigma_3)}{2}$.

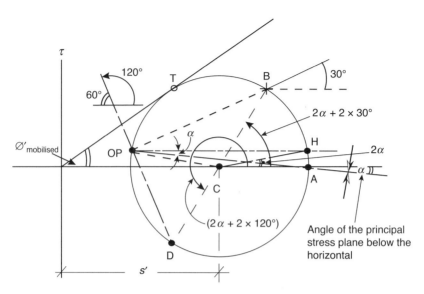

Figure 4.19 The Mohr circle of stress.

c) The orientation of the plane on which σ_1' acts α to the horizontal reference plane.

There are three unknowns s', t, α, but there are three known stresses at known angles, which yield three equations. Based on the fact that the angle subtended by a chord (e.g. AH) at the centre of a circle $\stackrel{\frown}{H\text{–}C\text{–}A} = 2\alpha$ the angle at the circumference $\stackrel{\frown}{H\text{–}OP\text{–}A}$.

The three equations can be written:

$$\sigma_{\text{vert}}' = s' + t\cos 2\alpha \tag{a}$$

$$\sigma_{30°}' = s' + t(\cos(2\alpha + 60°)) \tag{b}$$

$$\sigma_{120°}' = s' + t(\cos(2\alpha + 240°)) \tag{c}$$

Solving these equations will provide the values of s', t and α.

The Mohr circle of strain in Figure 4.20 was introduced in Chapter 3; it is again used here to determine the principal plane from a strain gauge rosette.

Again the Mohr circle of strain has three unknowns:

(a) The centre $= \tfrac{1}{2}$ volumetric strain $\frac{(\varepsilon_1 + \varepsilon_3)}{2} = \frac{\varepsilon_\nu}{2}$.

(b) The radius $= \tfrac{1}{2}$ the maximum shear strain $\frac{(\varepsilon_1 - \varepsilon_3)}{2} = \frac{\gamma_{\max}}{2}$.

(c) α the plane *across* which the major compressive strain ε_1 acts.

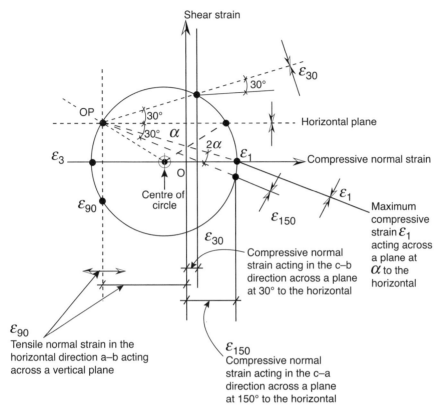

Figure 4.20 The Mohr circle of strain.
           Note: It must be remembered that normal strains are assumed to act *across* the plane at right angles to their alignment. See Figure 4.21.

We have measured three normal strains in known directions to the horizontal $\varepsilon_{90°}$, $\varepsilon_{30°}$ and $\varepsilon_{150°}$. Again, the planes *across* which they *act* are key, on the same basis as the stress circle, three equations can be written:

$$= \varepsilon_{0°} = \frac{\varepsilon_{v}}{2} + \left(\frac{\gamma_{max}}{2}\right)(\cos 2\alpha) \qquad \text{(a) – note } \varepsilon_{horiz} \text{ is tensile}$$

$$\text{Horizontal plane} \quad = \varepsilon_{30°} = \frac{\varepsilon_{v}}{2} + \left(\frac{\gamma_{max}}{2}\right)\cos(2\alpha + 60°) \quad \text{(b) – compressive}$$

$$= \varepsilon_{150°} = \frac{\varepsilon_{v}}{2} + \left(\frac{\gamma_{max}}{2}\right)\cos(60 - 2\alpha) \quad \text{(b) – compressive}$$

Note: there is an additional observation to enter here. If a material is elastic; the total stresses and the total strain coincide in the principal directions. Also the incremental

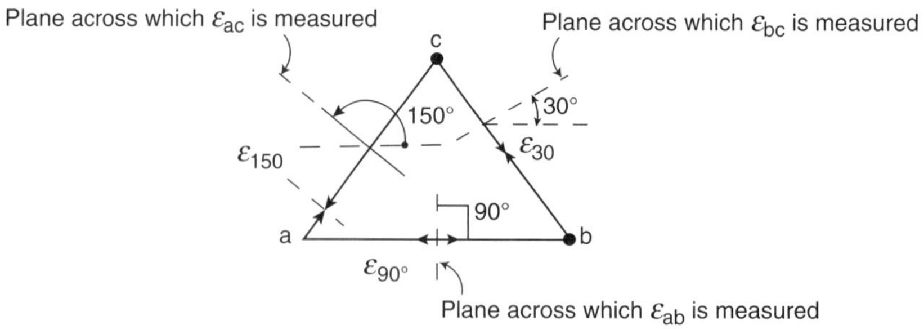

Plane across which ε_{ac} is measured   Plane across which ε_{bc} is measured

ε_{150} $150°$ $30°$ ε_{30}

$90°$

a $\varepsilon_{90°}$ b

Plane across which ε_{ab} is measured

Figure 4.21 Orientation of a strain gauge rosette and the planes across which nominal strain acts giving rise to revised nomenclature of the strains.

strains and the incremental stress directions coincide. *But* if the material is perfectly plastic the *incremental principal strain directions* will coincide with the *total principal* stress direction, *not* with the *incremental principal stress directions* (provided the *incremental stress* still results in *plastic deformation*).

If the behaviour is elasto-plastic, the σ_1', $\Delta\sigma_1'$ and ε_1 or $\Delta\varepsilon_1$ directions will have complex relationships, but can indicate the proportion of plastic to elastic strains that are developing.

Pore pressure change

The measurement of natural pore water pressure is much more reliable, provided, as said earlier, there is an intimate and responsive direct connection. The intimate connection must be ensured by 'good' installation practice. The responsive component depends on the permeability of the soil as it can take some time for pore water pressure changes within the soil, including those caused by the instrument installation, to reach stable equilibrium. The time can sometimes be weeks and deciding the *in situ* pressure before civil engineering work commences is often compromised by lack of pre-construction time.

Once equilibrium is established, pore pressure changes influence a very extensive zone, and quite small piezometers will be accurate and reliable, and data can be used to contour wide areas.

The data quality may be good, but interpretation of its meaning may be considerably more difficult. In Chapter 2, some basic soil mechanics theory was developed and an idealised, simplified model was introduced, and the development of pore water pressure changes were illustrated on the basis of this simplified model. There are in fact a number of more sophisticated models (e.g. the 3SKH model) which are being continuously developed and incorporated into various numerical codes. The reader can use the simple approach given in Chapter 2 to obtain some appreciation of the changes observed, but the author recommends that complex loading problems should be back-analysed numerically using the most up-to-date codes available.

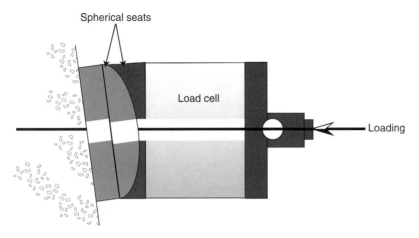

Figure 4.22 Typical load cell arrangement on a tension anchor.

Interpreting other vibrating wire instruments, load cells and crackmeters

Load cells are by their nature and design uniaxial items. Earlier it was explained how, for recording soil stress changes in two dimensions, three units at three known orientations could be used to determine the values of mean normal stress s_{total}, maximum shear t and the orientation of the principal stress plane. Otherwise they only measure the normal or axial stress across the active surface.

Axial loads in structures can only be usefully determined under individual columns, at the end of props, under bridge or beam seats and under the bearing plates of tension anchors. In all cases the single load direction is fixed and known.

It is surprisingly difficult, however, to provide a reaction plate for a load cell which is precisely at right angles to the direction of the force and can adjust if its direction changes slightly. In these circumstances a pair of hardened spherical seats should preferably be incorporated as in Figure 4.22.

Cracks and damage measurement

The development of cracks is essentially the response of brittle unreinforced materials to differential deformations or excessive temperature fluctuations. The reason is the very low tensile strength of the material in contrast to its compressive strength and compressive stiffness. Common materials in this category are bricks and mortar, masonry and (under unusual historic conditions) dried mud constructions. It must be remembered that the vast majority of domestic and historic buildings fall into this material category.

Cracks are initiated at points of high stress concentration, very often caused by the location of openings in the structure (such as entrances and windows), or by the inadequate design of foundations which results in differential settlement, or by external work (tunnelling, deep excavation or drainage) that causes differential settlement.

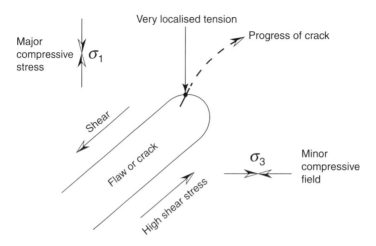

Figure 4.23 Progress of a Griffith's crack.

The commonest cause of crack initiation and development is the development of displacements that result in a direct tensile stress which exceeds the material capacity. An alternative is the development of a shear crack. This type of crack can develop even under two axes of compressive principal stress if there is some orientated stress raiser, and the reader is referred to any materials textbook on Griffith's crack propagation theory.

Figure 4.23 shows the end of an idealised flaw or an existing crack under σ_1 and σ_3 that are both compressive. The end of the crack is assumed to be a very small radius semicircle, the detailed stress distribution round this semicircular stress-free surface results in the development of a small zone of high tensile stress, and the crack propagates as shown.

Consider now two surprisingly common situations (Figures 4.24 and 4.25).

Block A has a widespread foundation area with integrity, whereas the brickwork column B between lines Y–Y and Z–Z is a concentrated load and will settle more at foundation level than A.

The windows become badly distorted and families of tension cracks open in the panels between windows; detailed shear failures can occur in the cills, pediments and their local bearing surfaces.

In later Victorian masonry and brick structures, particularly railways and heavy industrial buildings, there was a major improvement in foundation design, in particular the inverted brick arch considerably stiffened the basement and spread the dead load as seen in Figure 4.26.

This foundation design is considerably more tolerant to localised deformation, but will still suffer if external foundation settlement is induced. A critical situation is that discussed by Boscardin and Cording (1989) and is shown in Figure 4.27.

The base from a to d suffers a hogging bending but also the foundation is stretched: point a moving to a′, b to b′, etc. The structure as a whole may act as a cantilever beam with the zone a–n–l–m–d–a suffering tension cracking in the mortar and bricks as shown. In the vicinity of l the cracks are probably vertical due to the tensile stress

Figure 4.24 Typical Georgian facade (brick/stucco faced) with inadequate brick foundations.

being dominated by the bending. Again along a–b–c the stretching to a'–b'–c' will induce vertical cracks within the body of the wall inside the susceptible zone, the crack orientation will become more inclined, again openings will have a stress-raising influence and initiate localised cracks.

Cracks induced by shrinkage and temperature variations are usually simply tensile across the direction of shrinkage.

Joint monitoring is the deliberate installation of a specialist instrument across a planned construction detail. This detail will avoid the random location of shrinkage cracks and provide a location with the ability to monitor the block displacements. Most joints are designed to prevent displacements in at least one, sometimes two directions. So the monitoring is simply a one-dimensional movement across the joint. In the case of random cracks a decision must be made as to whether the cracks are simply a tensile displacement orthogonal to the crack, or mixed tensile and shear, producing displacements both orthogonal and parallel to the crack (some construction joints allow both

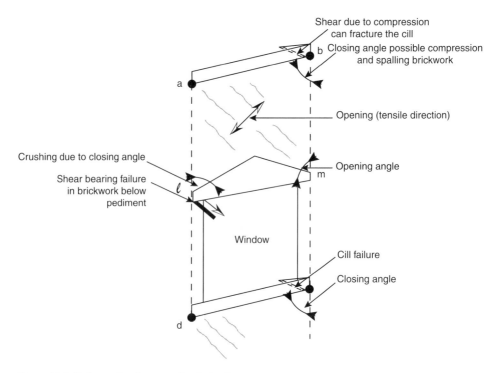

Figure 4.25 Deformation between loaded columns.

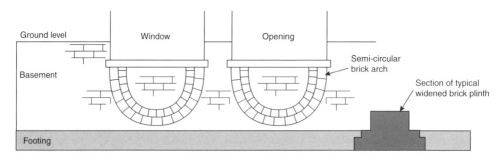

Figure 4.26 Inverted brick arches.

movements). In unusual circumstances (under high compressive stress conditions, rock caverns for instance), while shearing, the fractures can experience crushing of asperites (roughness) on the fracture surfaces significantly reducing the roughness coefficients, and effectively reducing the apparent friction capacity of the fracture. Such situations can fail explosively, and crack monitoring should be remote and include a rate of change assessment.

Finally, if the crack is self-generating within a wall or rock mass it can normally only be viewed on an exposed surface, and then the development of the three-dimensional

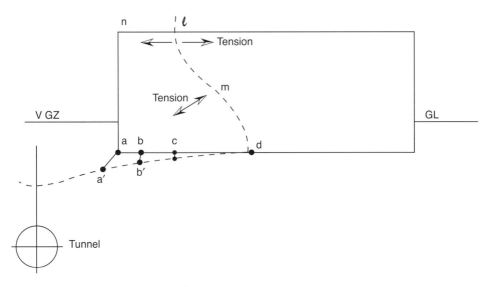

Figure 4.27 Differential movements due to tunnelling.

mechanism into the wall or structure may result in a block moving out of the observation plane as well as two movements within it. In this case all three directions of movement must be monitored (using three orthogonal directions).

Crack monitoring instruments

Essentially a crack will have a clear orientation on a surface and basic movements can be measured:

a) Orthogonal to the crack.
b) Orthogonal to the crack and parallel to the crack.
c) As b) above but also at 90° to the observational surface.

All setups require two *stable anchor* points, one each side of the crack. This may not actually be as simple as it first appears because cracks do not form as isolated single entities. Initially, a field of parallel orientated cracks forms in the area and eventually *one* will become dominant and carry all displacement as a localised movement. The result of the earlier field cracking is that anchor points can be in fractured material and not stable.

However, as the movements are assumed to be associated with the single location, the measurements are absolute displacement values, so the separation of the anchor points is irrelevant as the material between the anchor point and its side of the crack is assumed to be rigid. The orientation and measurements are shown for the three cases in Figures 4.28–4.30. Figure 4.31 shows the potential three-dimensional crack displacements in a poorly constructed concrete road deck.

Figures 4.32–4.34 show actual examples of Figures 4.29–4.31, respectively.

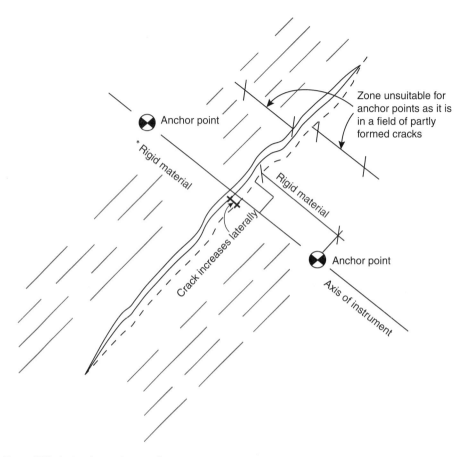

Zone unsuitable for anchor points as it is in a field of partly formed cracks

Anchor point

*Rigid material

Rigid material

Crack increases laterally

Anchor point

Axis of instrument

Figure 4.28 A simple tension crack monitor.

The measuring units are linear displacement transducers working on either vibrating wire technology (see this chapter) or simple, robust linear potentiometers (typically used in a half-bridge configuration). Normally the displacement ranges for these instruments are 50–100 mm, mainly due to the need to accommodate the separation of the anchor points. However, the author would point out that 10 mm or more of crack displacement in brick and masonry buildings is of major significance; if movements reach 50 mm, significant damage or even local failures will probably have already occurred.

The anchor units for monitoring planned joints are fundamentally identical. However, they are at a known location, with known orientation and there are no surrounding damaged areas. Units can therefore be installed during construction and both during and post construction displacements measured from a real zero (datum).

Situations in which joints are important are concrete structures such as slab structural walls in excess of 10 m to 15 m in length, concrete mass reinforced retaining walls,

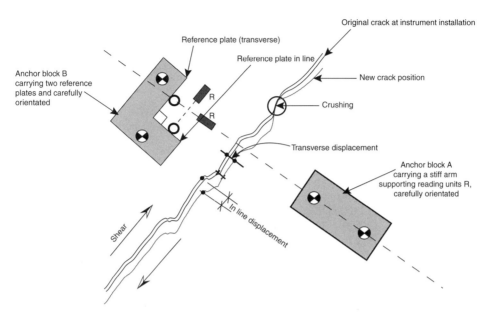

Figure 4.29 Shear and tensile crack monitoring.

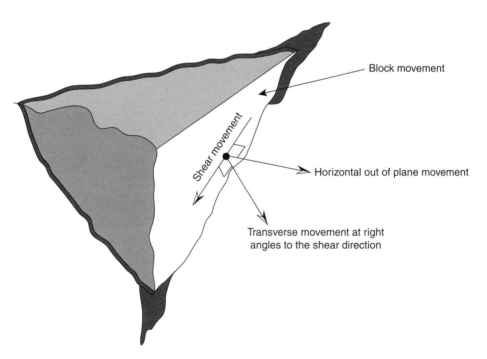

Figure 4.30 Three-dimensional crack monitoring – example A.

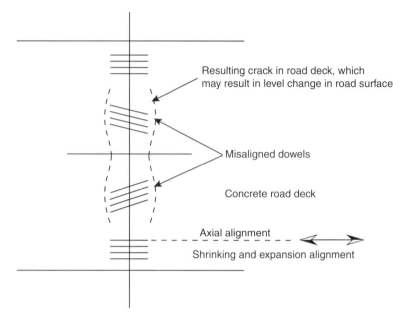

Figure 4.31 Three-dimensional crack monitoring – example B.

Figure 4.32 An example of simple tension crack monitoring.

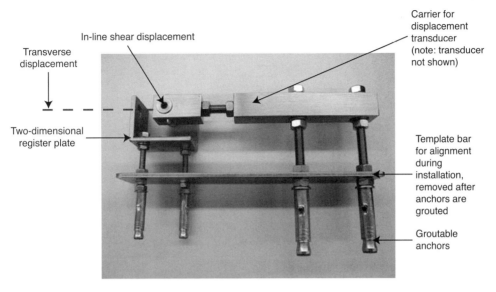

Figure 4.33 An example of transverse and aligned displacement crack monitoring.

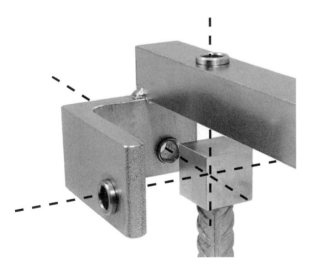

Figure 4.34 An example of a three-dimensional crack meter.

concrete dam construction, concrete built as stepped towers, or upstream membranes in panels and concrete road and rail decks.

It is important to know what sealing techniques were incorporated in the joints as this may (rather than structural distress or failure limits) control alarm thresholds for the allowed movements.

Chapter 5

Survey techniques

Total station monitoring

Theodolites have, for more than 200 years, been a fundamental instrument of surveying – improving to become a one second of arc or better instrument – along with the chain/tape and level. In the mid 1960s a new unit, the EDM (electronic distance meter) became available and revolutionised distance measurement by utilising the phase difference of infrared light transmitted and received from a reflector at the monitored location. The emission accuracy is now to within 0.5 mm.

This precise distance measuring revolutionised surveying when incorporated into the telescope of a theodolite. From a levelled, orientated and referenced station the theodolite action enabled:

a) horizontal angle with respect to North θ_N° to be recorded;
b) the elevation angle with respect to the horizontal δ° to be measured, both to 1 second of arc; and
c) the EDM unit to give the value d and hence the height and the horizontal distance measurements to $\pm$ 1 to 3 mm.

All this required an experienced surveyor at the instrument.

The further development of low-cost computing power, higher-precision measuring devices within the instruments and precision controlled stepper motors in the 1980s and 1990s brought about a second revolution; combined EDM/theodolites were motorised and self-reading, the computer control enabling the unit to search for a monitored point based on the previous known location and then read the current data $\theta_{n.t}\delta_{n.t}d_{n.t}$ at time t.

These units were not cheap as a capital investment and were still accurate to just 1 second of arc. Currently some expensive high-accuracy units measuring to 0.1 seconds of arc are available. Figure 5.1 shows a typical modern unit, termed a 'total station'.

Use of automatic total stations

The fundamental operation of an automatic total station (ATS) is to locate a monitored point with respect to the ATS master location. This master location is defined in terms of three physical positions; Northing N_0, Easting E_0 and height above datum H_0 and

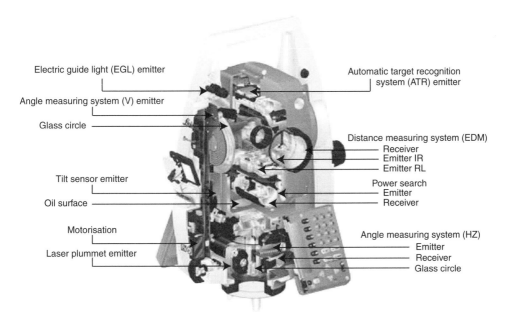

Figure 5.1 Typical modern 'total station' (courtesy Leica Geosystems).

the orientation of the site North (N↑). The monitored point n will then be defined by three pieces of data $\theta_{n.t}\delta_{n.t}d_{n.t}$ the horizontal angle from site North, the elevation angle and the straight line (slope) distance to n, all at time t.

Computation is simple (see Figures 5.2–5.5) the value of $\delta_{n.t}$ and $d_{n.t}$ give the key horizontal distance $d_{h.n.t}$ and the vertical difference $h_{n.t}$. The $N_{n.t}$ and $E_{n.t}$ values are simply computed horizontal offsets based on the calculated value of $d_{h.n.t}$ and $\theta_{n.t}$.

This is the simple part of the use of a total station, where the instrument is located on an immovable, referenced location (for example an Ordnance Survey trig. point). In field instrumentation this situation is extremely unlikely and the total station's mounting may itself be subject to displacements due to the engineering works. If this is the case then corrections at time t to the N_0, E_0, H_0 and N↑ reference of the measuring instrument must be added to the calculation.

This correction is conventionally carried out by back-reference to a number of 'fixed', stable locations, which are considered to be outside the influence zone of any

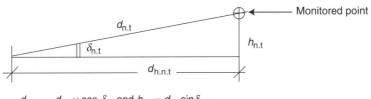

$$d_{h.n.t} = d_{n.t} \times \cos \delta_{n.t} \text{ and } h_{n.t} = d_{n.t} \sin \delta_{n.t}$$

Figure 5.2 Vertical section of a sight line.

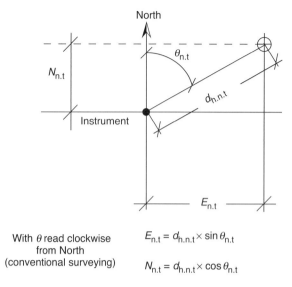

$$E_{n.t} = d_{h.n.t} \times \sin\theta_{n.t}$$

With θ read clockwise
from North
(conventional surveying)

$$N_{n.t} = d_{h.n.t} \times \cos\theta_{n.t}$$

Figure 5.3 Plan of a sight line.

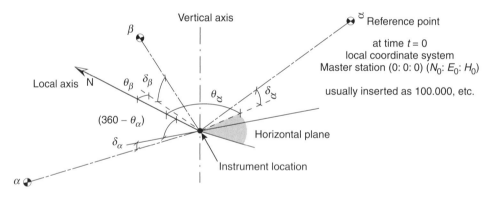

Figure 5.4 Isometric drawing of the reference points.

assumed engineering work. However, as every building is subject to daily, monthly, seasonal and yearly cyclic movements, the choice of these fixed locations must be made with considerable care. They should be on or near the foundations of major structures and they should preferably be in the shade and not subject to external vibration. Figure 5.4 shows an isometric view of three fixed points α, β, and $\eth$ (note the 'spread' should preferably cover a 120° angle horizontally and possibly a 15° to 20° angle vertically). Figure 5.5 shows this in a plan view.

The procedure is to assume only localised 'relative' displacements are required, so at time $t = 0$ the ATS is set at a convenient set of coordinates (typically N, E and H all $=$ 1000.00) and the horizontal angles are set to site N (at 000°00′00″). The three reference points are treated like monitoring points and their N_α, E_α, H_α, etc.) are calculated and

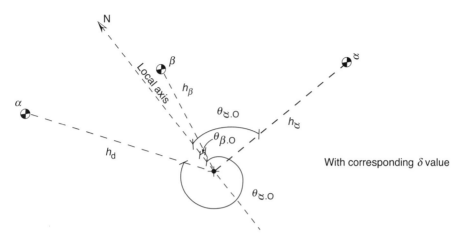

Figure 5.5 Plan view of the reference points.

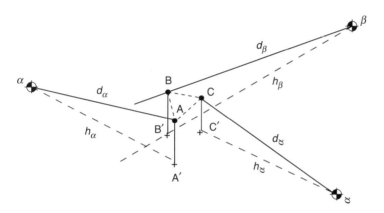

Figure 5.6 Isometric diagram of reciprocal sights from reference stations.

assumed to retain those values throughout the project. Hence these coordinates are now the assumed stable reference coordinates for α, β and $\eth$, the computing data base. The locations of α, β and $\eth$ are shown isometrically with three new sightings at time t in Figure 5.6 and in plan in Figure 5.7, with readings to the reference stations drawn as reciprocal sight lines.

There are in effect nine key pieces of information, the three horizontal and three vertical angles (which are to one second of arc) and the three horizontal distances which are probably to ±2 mm. This locates apparently three positions of the ATS unit as shown by A, B and C in Figure 5.6 and by A′, B′ and C′ in Figure 5.7. This error triangle can first be reduced to a minimum by rotating the reference N axis by the value of $\Delta\theta_{\text{n.t}}$ as shown in Figure 5.8 (being adjusted anticlockwise in the illustration by δ_n° to provide locations A″, B″ and C″). Adding, for example, the error

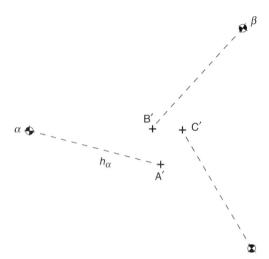

Figure 5.7 Plan view of Figure 5.6.

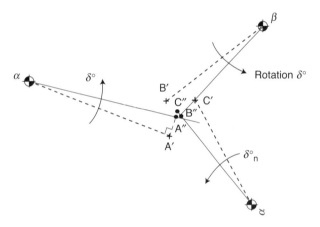

Figure 5.8 Correction for North reference angle.

bar of ± 2 mm on the distance measurement gives Figure 5.9 and the best-fit location for the ATS at point X.

Other adjustment methods are available, the corrections are usually small as the location chosen for the ATS will be the most stable available. More than three reference points can be backsighted, in which case all data are weighted and a least squares calculation used to identify the corrected coordinates of the ATS at time t.

There is now a best-fit position for the master station at time t in terms of $(N_t, E_t, H_t$ and $N_0 + \Delta\theta_{nt})$ as shown in Figure 5.10.

Using the adjusted position and the corrected distances there will now have to be a correction to the height ΔH_t. The new master station distances to the three reference points and the original measured elevation angles are used to back-calculate three

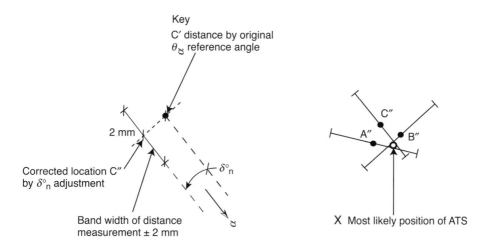

Figure 5.9 Inclusion of distance error band.

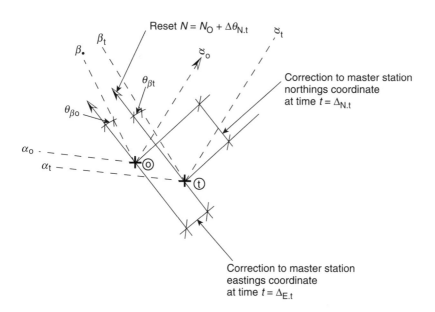

Note there will also be a correction to the height $\Delta_{H.t}$ and to the N reference angle.

Figure 5.10 Corrected northings and eastings of the ATS.

height values for the master station (see Figure 5.6), the mean value is now used as the H_t in subsequent calculations.

The positions of each general monitoring point (θ_{xt}, δ_{xt} and d_{xt}) are taken followed by a recheck of the reference stations. The local coordinates N_{xt}, E_{xt} and H_{xt} of each monitoring point are calculated and stored together with the start and finish master station coordinates. Check limits must be preset which if exceeded call for an automatic rerun of the measurement cycle.

Under good civil engineering site working conditions this form of determination of the three spatial coordinates at any moment t is probably repeatable and accurate to approximately ±2 mm. If the very accurate ATS (0.1 second of arc on angle measurement, 0.5 mm in distance measurement) is used, then ±1 to 1.5 mm may be achieved. However, difficult optical conditions such as rain, heat haze, dust, strong winds, machinery exhausts, etc., can cause serious degradation of measurement accuracy.

Having said this, it must be remembered that the ATS is the only technique currently available that can provide all three pieces of spatial data for visible points on civil works, the most useful aspect of which is local relative movement.

Key situations can be the deformation of major facades of buildings, e.g. St Pancras Station, London, during local tunnelling work, see Figure 5.11.

Another typical use is to monitor the displacements of old railway gravity retaining walls when the track bed is excavated and lowered during overhead electrification, and specifically the displacement of existing railway tracks when subjected to traditional and jack tunnelling with minimal crown cover, see examples in Figures 5.12 and 5.13.

Figure 5.11 The facade of St Pancras Station in London monitored as part of a new London Underground ticket hall excavation below it.

Figure 5.12 Traditional tunnelling beneath a live railway track.

Prisms

The monitoring point is derived from the original EDM system, the infra-red light beam is transmitted from the ATS as a near single beam of modulated light and is reflected via a series of three reflective surfaces that form a corner of a cube, which when viewed down the axis of sight appear as a 120° set of surfaces. This ensures that the light retraces its original path back to the ATS origin (now acting as a receiver). These reflecting units need to be as efficient as possible, as there are considerable losses over distance, and distortions will always try to bend the beam of light and its subsequent reflection in a non-advantageous manner. The current most commonly used prism targets consist of three precision ground surfaces intersecting as in Figure 5.14, metal back-coated to minimise light loss and front-coated to minimise front glass surface reflections (both internal and external). A diagrammatic representation of this can be seen in Figure 5.14

One rather simplified and exaggerated light path is shown, first hitting the front surface a at $\alpha°$ to the axis alignment, then refracted down to b, across to c and subsequently to d and exiting at e parallel to the incoming beam. The prisms are special-quality glass to ensure high performance and uniform refractive index. As the refractive index is 1.5 compared with air, the distance recorded appears to be slightly greater than it should be, see Figure 5.15. Also the apex of the prism is apparently moved (see Figure 5.16) due to poor face orientation.

Figure 5.13 Jack box tunnelling beneath a working station.

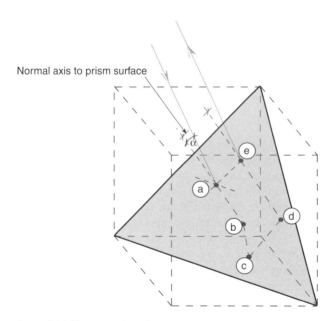

Normal axis to prism surface

Figure 5.14 Diagram of a reflecting prism.

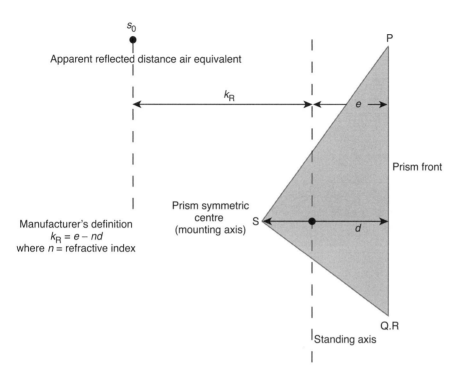

Figure 5.15 Cross-section through a triple prism.

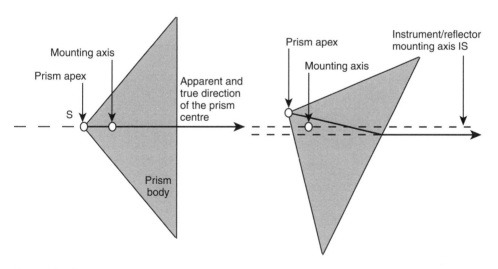

Figure 5.16 True and apparent recognition of prism centres.

The prisms are commercially available from specialist manufacturers, and a typical unit is shown in Figure 5.17. Units work most efficiently and accurately within ± 10° orientation to the instrument direction – see Figure 5.18.

The prism needs to be attached to the key points on the monitored structure so as to be nearly (±5° to 10°) normal to the line of sight from the ATS. In very difficult locations this angle may be increased, but the reflected signal will be weakened and poor atmospheric conditions may well preclude readings.

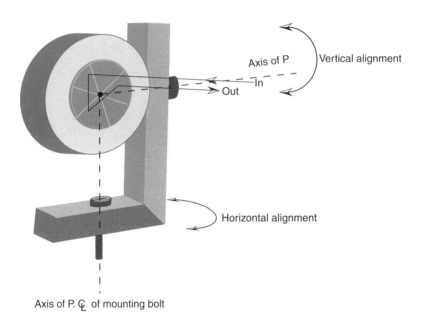

Figure 5.17 Prism showing the light path.

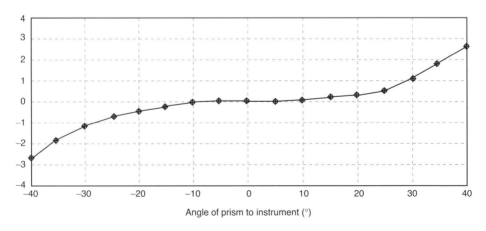

Figure 5.18 Deviation of the axis in dependence of the misalignment.

Figure 5.19 Typical prism in mount attached to a wall.

Mounting systems

Normally, the bracket on the prism shown in Figure 5.19 will enable the reflector to be fitted and adjusted to any stable structure. The units should not be fitted to loose bricks or damaged surfaces.

For use on railway lines, there are a number of serious restrictions that need to be taken into consideration. Initially it was forbidden to attach prisms to the rails and therefore they were glued or screwed to the wooden or concrete sleepers. This proved in several cases a poor idea as the sleeper's additional displacements often do not fully reflect the movement of the rail (which is the part of key importance), and particularly in the case of glue mountings (where drilling the sleeper was not allowed), these were easily damaged by vibrations or track maintenance.

An adapted pandrol clip mounting has now been accepted as the most effective way of coupling a prism directly to the underside of a rail, and is shown in Figure 5.20.

Figure 5.21 shows schematically a series of monitoring points on a piece of track and the resulting displacements from zero to time t. The data can provide information on slew (lateral movement), settlement profile (overall vertical movement) and most significant differential cants (twist or tilt).

Limitations in tunnels

Particular care must be taken when using an ATS system in railway tunnels as a number of issues, mostly relating to the confined space available, become relevant. Of these, the two most important are: the positioning of any backsight targets to allow for a resection to correct any movement of the ATS, and what are known as multiple prism errors.

Figure 5.20 A prism mounted on a pandrol clip.

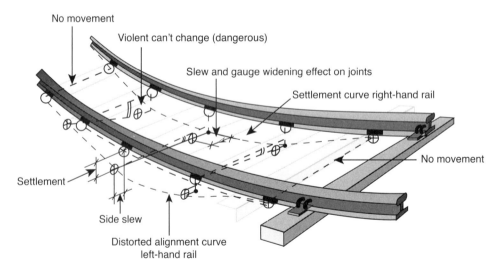

Figure 5.21 Track monitoring (note: track distortion exaggerated).

Figure 5.6 showed an ideal layout for backsight targets where they are located behind the instrument on stable locations in an approximately 120° horizontal and 10° to 15° vertical 'spread'. In a tunnel, due to the need to have targets at section locations within the sight of the ATS (and therefore necessitating their installation in near-straight sections of tunnel), the horizontal angle change between the targets is small and becomes smaller the further from the ATS they are placed, leading to a set-up which is far from the ideal, Figure 5.22. Further complications arise when the backsight targets are so close to the ATS that they are in the zone of influence and thus cannot be used for effective resection correction.

Similar physical restriction problems exist for multiple monitoring targets. If, for example, a series of arrays consisting at each section of prism targets (viewed clockwise from the ATS) are installed at trackbed level, knee, shoulder, crown, shoulder, knee and trackbed (i.e. forming an array of seven targets around the lining at any given chainage – see Figure 5.22), then if one prism is observed in array number 1 (say the knee) then the same prism in arrays 2, 3, 4, etc., may well be visible separately on the left side, but in the field of view provided by a modern ATS, those on the right side will result in multiple returned light paths. Resolution of the data will be impossible.

This can be overcome by using two instruments on either side of the tunnel and by, where possible, 'staggering' prisms around each array so they form a 'helix' pattern when viewed longitudinally; however, both solutions are time consuming and expensive to install.

Recent ATS developments go some way towards negating the multiple prism errors, but even so, great care must be taken when designing and installing an ATS system to overcome this fundamental problem in tunnels.

Other areas which require careful consideration are the effects of temperature stratification (particularly along the side walls), vibration from passing trains, prism cleaning (which can need to be monthly in highly trafficked tunnels) and electromagnetic interference from DC power supplies.

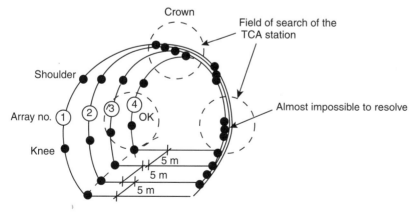

Figure 5.22 View from the ATS station of four arrays of reflections showing multiple prism interference.

Recent work in the United States has incorporated a laser unit in a total station to monitor reflectors, it has also been able to operate it without the reflector. This has enabled areas such as road surfaces which are in use to be monitored by fixing the sight line for repeat readings. Any variations in laser distance can be used to monitor the level changes.

More advanced optical scanning will now be discussed.

Advanced optical systems

There are other optical systems which can be used to determine deformations, the more advanced of these are discussed below.

Photogrammetry

The use of two photographs to provide a stereo image of a local surface has been employed since the 1930s and enables a large area of land to be contoured. One of the amusing features of the technique was that of the 'flying sheep'. As the aircraft noise disturbed the sheep between the first and second photograph they appeared to 'take off' from the ground! If two identical photographs are taken from the same place at different times then anything that has moved will stand out (i.e. appear to fly) in the same way. This apparent height change can be measured by an operator based on a parallax calculation. With the advent of accurate digital cameras (which are becoming progressively cheaper) it is now possible to use three or more cameras at permanent locations, or to use multiple photographs taken from a helicopter for instance, and produce a three-dimensional image of the photographed face or of the whole structure. A similar set of photographs taken at a later time can be compared to the initial analysis, and any displacement quantified with respect to the stationary areas within the photographs.

As the important deformations are relative within the structure or the surrounding area, any movement of the location of the cameras can be adjusted for by finding the statistical best fit from all the key pixel data of the whole structure and its surroundings, and then examining the displacements that differ from the statistical fit. If the static cameras are within 15 m of a facade being monitored, an accuracy of better than 1:100,000 pixels can be achieved with some systems reporting up to 1:200,000.

With the multi-photographic images from a moving source (such as a helicopter) a similar best fit can be located, the position of the cameras for each photograph being determined, and again the three-dimensional building details computed for a statistical best fit. A second complete sweep of photographs will identify areas deviating from the initial information.

Laser scanner systems

The development of accurate laser sources and advanced reflectorless measurement enables the accurate measurement of light travel times, and no longer requires any phase shift assessment as used in EDM units and fibre-optics.

Lasers fired at any surface will provide some reflection, and this reflection can be received by the instrument and the return time very accurately measured. This is a similar principle to radar, but being based on a single frequency light wave is considerably more accurate. The Lidar system adopts this principle for terrain mapping. The system is used from aircraft or helicopters flying along several tracks and can cover large areas.

The response from flat surfaces is very distinctive and the software initially searches for roads, paved areas and roofs. The lateral sweeps are then used to identify walls and other solid objects. Finally the cloud of laser return points is used to assess the open surrounding landscape; reflections from trees and crops can be complicated. The data take a considerable amount of complex software to analyse and provides block models of buildings, ground contours and point details.

If conventional digital aerial survey photographs are taken at the same time, then these can be numerically superimposed on the block and terrain model to provide virtual reality. A second later survey can be used to determine deformation changes.

Its accuracy is in the order of $\pm$ 5 to 10 cm and it is therefore useful for monitoring cliff and steep slope instabilities where continuous movements are occurring and personnel cannot or should not have access to the moving area.

A more-sophisticated system based on a laser instrument can be used from ground-mounted units set-up in a similar way to a total station to monitor structures within 150 m (e.g. Leica Cyrax) and works as follows.

The laser is fired at a precision rotating mirror set at 45° to the light source. This mirror spins at high speed and sends out a fan of laser pulses, diagrammatically depicted in Figure 5.23. These are reflected and the return time and trajectory measured.

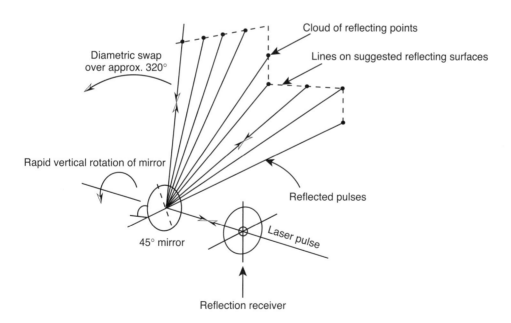

Figure 5.23 Diagrammatic representation of the reflected locations for a single sweep.

At the same time the body of the unit rotates slowly on a horizontal axis. The result is a spherical fan of light pulses producing a cloud of reflection points. For every point covered by this sweep (and there may well be hundreds of thousands of them) the system records: the horizontal angle, the orientation of the mirror (giving the vertical angle) and the distance to each point, i.e. the reflecting object. Fundamentally, this is total station data, except that millions of points are measured within a few minutes.

A further set of data from another fixed location is combined with this initial set and the combined system is statistically merged. The required software and data storage are massive and complex; initially the system again seeks distinct near-planar areas such as roofs, walls, windows, etc., but in more detailed form (Figure 5.24).

An overall three-dimensional solid block image is produced of the building front. A photographic sweep can be applied from above or in the place of the laser scanner. The visual images are then distortionally adjusted and laid over the three-dimensional block images to provide a virtual image of the face of the structure. Key points can be identified and given x, y and z coordinates. A beneficial addition to this general reflected image is to incorporate prisms, at this close range they provide strong point reflections which can be used as key references when processing the images. They can also be analysed more rapidly and quickly provide deformation indications within minutes of the second scan, while the full cloud analysis and the virtual image manipulation may take rather longer.

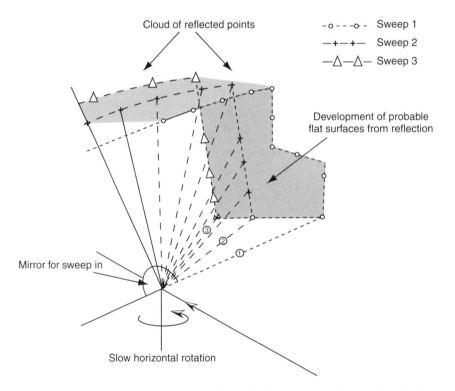

Figure 5.24 Diagrammatic representation of the development of surfaces with multiple sweeps.

Figure 5.25 The Faro Photon 120/20 laser scanner.

If the system can be restricted to a field of approx 150 m radius the accuracy of individual points can be as high as 0.3 mm to 0.1 mm. If a second scan is conducted from the same locations at a later date then relative movements can again be determined to better than 0.5 mm. This level of accuracy is now well within the required range for structural monitoring associated with tunnelling, for instance the unit (e.g. Figure 5.25) can be used from any number of locations that are fitted with a total station tribrach.

Chapter 6

Specialist instruments

The conventional manual convergence measurement system is described in Chapter 7 (page 167) this chapter describes an automated version that enables readings to be obtained when human access is not possible or where automatic readings are required.

The Bassett convergence system

A convergence system requires the ability to determine the Δx and Δy displacements at a number of reference points in a section. The electrolevel and microelectromechanical systems (MEMS) measuring units are as described earlier (Chapter 3). These two units can achieve both considerable resolution of small angle changes, the electrolevels for small angle changes (a typical near-linear range of 40 arc minutes) and the MEMS with less accuracy but over quite wide ranges (typically 10° or more).

The Bassett convergence system (BCS) has been designed with attention to the above performance criteria. The resulting system is described below.

The basic building block of the BCS is the bar/cam unit. Various key components are used to assemble the bar/cam unit, they comprise:

1) Reference pin.
2) Collars.
3) Long bar/cam coupler.
4) Cam pin.
5) Electrolevel sensor unit.
6) Cam MEMS unit and cam mounting.
7) Long bar (straight or curved).

Any number of these bar/cam units may be interlinked to provide a system of the required size and shape. The units can be assembled in a number of configurations to cope with obstructions or peculiarities of the reference line. Figures 6.1–6.3 show the key components.

The geometric principles of a bar/cam unit

The bar/cam unit can be used to provide Δx and Δy displacements between two fixed points on a structure. The process by which this is achieved is illustrated in

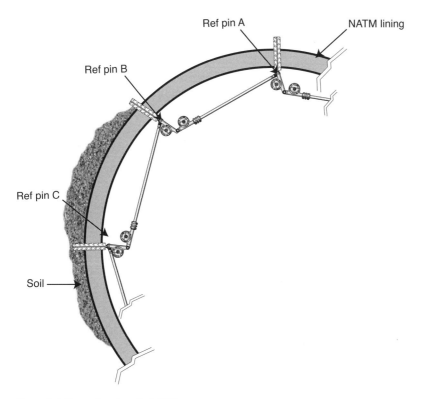

Figure 6.1 Part of an installed BCS.

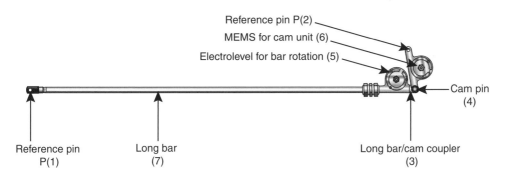

Figure 6.2 Components of the BCS.

Figures 6.4 and 6.5 by considering the displacement between two reference pins P_1 and P_2 (Figure 6.4). After initial determination of the base lengths of ℓ_{0x} and ℓ_{0y}, $\Delta \ell_0$ and $\Delta \alpha$ are monitored or determined (Figure 6.5). Using local coordinates this will provide Δt and $\Delta \ell$ (Figure 6.5), and using the angle $A°$ (Figure 6.4) can provide horizontal and vertical displacements Δx and Δy (Figure 6.5).

Figure 6.3 BCS with bent long bar.

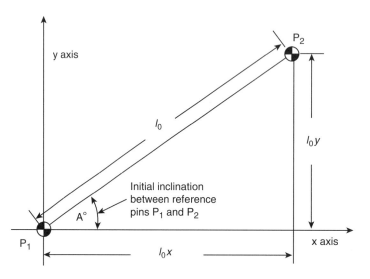

Figure 6.4 Reference pin.

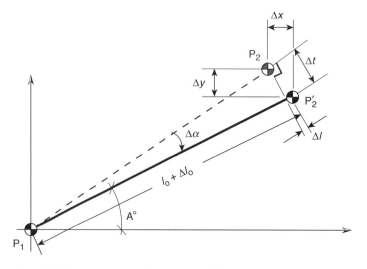

Figure 6.5 Basic rotation and extension to P_2.

A simple arrangement would be to directly measure $\Delta\alpha$ in Figure 6.5 with an electrolevel and $\Delta\ell$ by an linear variable differential transformer (LVDT) aligned precisely with the P_1–P_2 direction. Difficulties occur with systems using LVDTs when the connecting line P_1–P_2 is obstructed and when high-energy electric discharges are present, making this approach impracticable and unsuitable. It is therefore necessary and desirable to use electrolevels and MEMS sensors to provide the values required to calculate Δx and Δy. How this can be achieved is best illustrated by considering the bar/cam unit in its basic form, i.e. a connecting beam B and a cam C (Figure 6.6) connected with precision bearings at the reference pins and at P_C. The cam C_0 is of a fixed length ($C_0 = 100\,\text{mm}$) and carries a MEMS unit, the beam B is cut to a convenient length in the field to suit the length (ℓ_0) P_1–P_2, and if possible to make the angle P_1–P_C–P_2 a right angle, though this is not absolutely necessary.

The action of the system is shown in Figure 6.7, the displacement of P_2 to P_2'' being accompanied by two angle changes $\Delta\alpha$ and $\Delta\theta$ in the beam and the cam, respectively. The geometry of this system is similar to valve gear linkages, and the full mathematics for angle X–P_c–P_2 (Figure 6.7), equal to β are set out below. The simplified form if X–P_c–$P_2 = 90°$ is also given.

Geometry of the beam/cam system

In Figure 6.7 movement of point P_2 to P_2'' has been divided into two separate rotations:

a) In Figure 6.8 the beam/cam unit has been rotated by an angle $\Delta\alpha$ about P_1 with the angle X–P_c–$P_2 = \beta$ kept constant. The unit is in effect a rigid link, P_2 therefore moves to P_2' sweeping a circular arc, while P_C moves to P_C' on a concentric arc. In terms of local coordinates $\Delta\ell_\alpha$ aligned along P_1–P_2 and Δt_α transverse to P_1–P_2. These resulting displacements are shown in Figure 6.9 and are

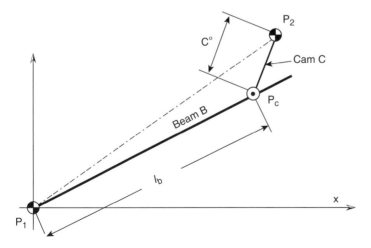

Figure 6.6 Insertion of the cam C to P_1–P_2.

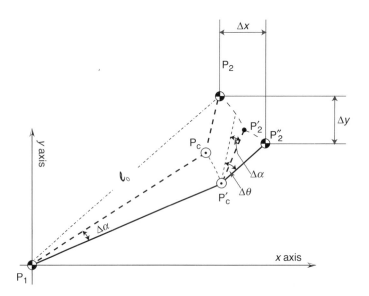

Figure 6.7 Action of BCS from P_2 to P_2''.

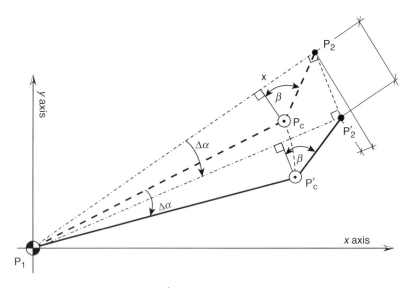

Figure 6.8 Key angles for P_2 to P_2'.

given by:

$$\Delta\ell_\alpha = -\ell_0(1 - \cos\Delta\alpha) \tag{1}$$

$$\Delta t_\alpha = \ell_0 \sin\Delta\alpha \tag{2}$$

It should be noted that with the beam/cam angle fixed at β the cam $(P_c' - P_2')$ has also experienced a rotation $\Delta\alpha$.

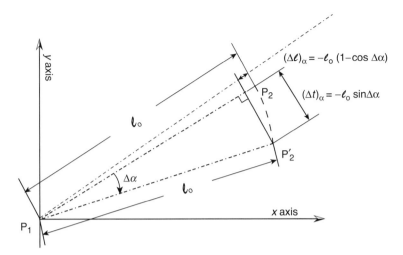

Figure 6.9 Displacements P_2 to P'_2.

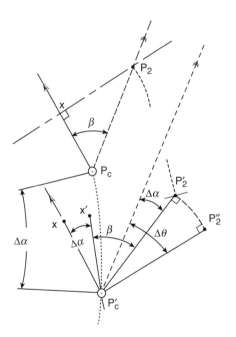

Figure 6.10 Rotation of the cam from P'_2 to P''_2.

b) Figures 6.10 and 6.11 show that as the cam unit is allowed to rotate, the measured rotation $\Delta\theta$ will actually contain the rigid rotation $\Delta\alpha$ as noted above. The reference point moves from P'_2 to P''_2. Again in terms of local coordinates (see Figure 6.11) along the original P_1–P_2 axis the additional displacements

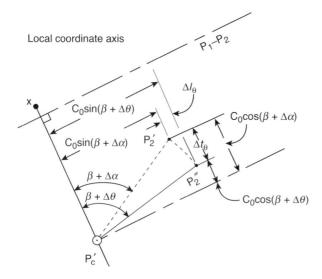

Figure 6.11 Displacement from P_2' to P_2''.

$(\Delta\ell_\theta)$ and (Δt_θ) are:

$$\Delta\ell_\theta = C_o(\sin(\beta + \Delta\theta) - \sin(\beta + \Delta\alpha)) \tag{3}$$

$$\Delta t_\theta = C_o(\cos(\beta + \Delta\alpha) - \cos(\beta + \Delta\theta)) \tag{4}$$

The full movement P_2 to P_2'' is therefore (see Figure 6.12):

$$\Delta\ell = -\ell_0(1 - \cos\Delta\alpha) + C_o\{\sin(\beta + \Delta\theta) - \sin(\beta + \Delta\alpha)\} \tag{5}$$

$$\Delta t = \ell_0\sin\Delta\alpha + C_o\{\cos(\beta + \Delta\alpha) - \cos(\beta + \Delta\theta)\} \tag{6}$$

rewriting (5) and (6),

$$\Delta\ell = -\ell_0(1 - \cos\Delta\alpha) + C_o\{\sin\beta(\cos\Delta\theta - \cos\Delta\alpha) + \cos\beta(\sin\Delta\theta - \sin\Delta\alpha)\} \tag{5b}$$

$$\Delta t = \ell_0\sin\Delta\alpha + C_o\{\cos\beta(\cos\Delta\alpha - \cos\Delta\theta) + \sin\beta(\sin\Delta\theta - \sin\Delta\alpha)\} \tag{6b}$$

Now if $\Delta\alpha$ and $\Delta\theta$ are small:

$$\sin\Delta\alpha = \Delta\alpha \quad \sin\Delta\theta = \Delta\theta$$

$$\cos\Delta\alpha \approx 1 \quad \cos\Delta\theta \approx 1$$

the equations simplify to:

$$\Delta\ell = 0 + C_o\{\sin\beta(0) + \cos\beta(\Delta\theta - \theta\alpha)\}$$

$$\Delta\ell = C_o\cos\beta(\Delta\theta - \Delta\alpha) \tag{7}$$

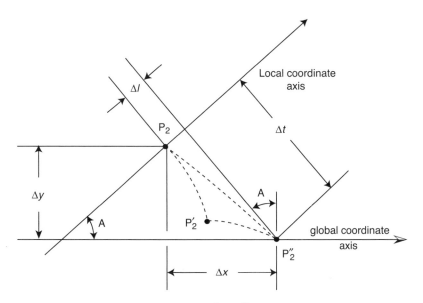

Figure 6.12 Summed displacement from P'_2 to P''_2.

$$\Delta t = \ell_o \Delta\alpha + C_o\{\cos\beta(0) + \sin\beta(\Delta\theta - \Delta\alpha)\}$$
$$\Delta t = \ell_o \Delta\alpha + C_o \sin\beta(\Delta\theta - \Delta\alpha) \tag{8}$$

Further, if β is set to zero degrees, the equations reduce to:

$$\Delta\ell = C_o(\Delta\theta - \Delta\alpha) \tag{9}$$
$$\Delta t = \ell_o \Delta\alpha \tag{10}$$

These are the local coordinate displacements based on an orthogonal system associated with the line P_1–P_2. With respect to a set of global displacements based on an x:y coordinate system the alignment of P_1–P_2 must be known, i.e. angle A in Figures 6.4 and 6.5 and the displacements Δx and Δy are simply:

$$\Delta x = \Delta t \sin A + \Delta\ell \cos A \tag{11}$$
$$\Delta y = \Delta t \cos A - \Delta\ell \sin A \tag{12}$$

In practice therefore, a datum P_0 must be chosen and the movements of all points P_0 to P_N determined by the summation of local Δx:Δy movements, and an initial precise survey is required to determine the initial parameters ℓ_0 and angle A and determined angle β, thereafter electrolevel changes $\Delta\alpha$ and $\Delta\theta$ determine subsequent relative and hence overall displacements. If a unit can be connected across the invert of the tunnel A to ℓ, then a closed traverse exists and any errors resulting from the summation of relative displacements can be distributed in the manner of a Bowditch correction.

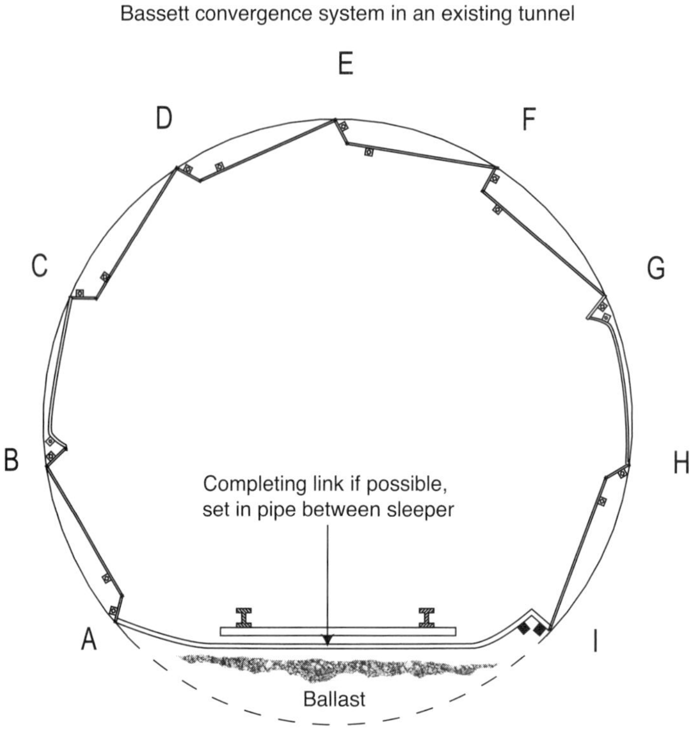

Bassett convergence system in an existing tunnel

Figure 6.13 Completely closed BCS.

There is no necessity for the link bar P_1–P_c to be straight as long as it is stiff, so it can be deformed to avoid any obstruction, as shown in Figure 6.3.

An example of the combination of bar/cam units shown in Figures 6.1–6.3, using the geometric principles described above for the monitoring of a tunnel in the Δx and Δy plane is shown in Figure 6.13. In this example the full ring has been closed by a deformed bar set below the track level linking A and ℓ. This enables the closure errors to be calculated and distributed. The bar will however be rather long and will almost certainly need damping to prevent the air pressure changes and vibrations from passing trains causing erroneous readings.

The BCS field trial, system configuration

The BCS configuration shown in Figure 6.13 was installed in a 3.85 m diameter cast iron tunnel lining of a London Underground running tunnel. Ten reference points were fixed to longitudinal flange bolt locations. A requirement of installation in an active running tunnel is that the system has a minimal intrusion into the tunnel and is formed to avoid interference with services in the sidewalls, and rails in the invert. The BCS can easily achieve this as shown in Figures 6.14 and 6.15.

Figure 6.14 An installed BCS system.

Figure 6.15 An installed BCS system (detail view).

Data capture and processing

Data were captured by a battery-powered Campbell datalogger, with communication via a modem link to a secure and safe location in a station some 70 m from the logger. Although not required in this trial, this communication system could easily have been accessed via the telephone network, thereby avoiding visits to the station except to change the battery every four months. Battery changing could also have been avoided by using a trickle charger off the tunnel utilities.

Frequency of data reading could be selected from a minimum of 30 s upwards, and programmed to vary with time. The communication, programming of the datalogger and processing of the data was carried out using a purpose-written program.

Figure 6.16 is a cross-sectional plot of the processed data, for one moment in time, from a large concrete-lined tunnel subjected to adjacent blasted construction. The reference points with magnified vectors of displacement are displayed.

Figure 6.17 shows the normal vector displacements of one individual monitoring point with time in a long-established SGI (spheroidal graphite iron) lined tunnel. Figure 6.18 shows the extreme displacements of this lined rock tunnel subjected to an unexpected adjacent blast (on the right-hand-side). Figure 6.19 shows the time movements of an individual monitoring point before, during and after this blast clearly showing the resulting permanent offset deformation. Figure 6.20 shows the data from Figure 6.19 but adds the permanent offsets experienced during previous blasts. The unexpected blast resulted in some three times the normal permanent displacement.

Measurements

From installation and commissioning in July 1994 to August 1995 the BCS trial system in the London Underground provided continuous, reliable and sensitive monitoring

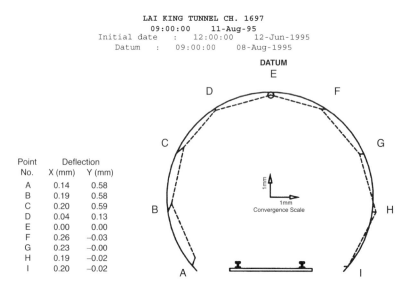

Figure 6.16 Data from a BCS system – cross-sectional plot of processed data.

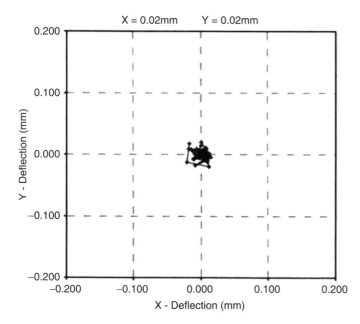

Figure 6.17 Data from a BCS system – normal vector displacements of an individual monitoring point.

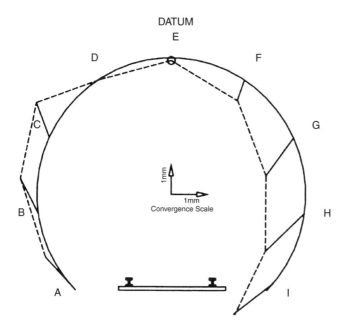

Figure 6.18 Data from a BCS system – extreme displacements of a lined rock tunnel subjected to an adjacent blast (on the right-hand-side).

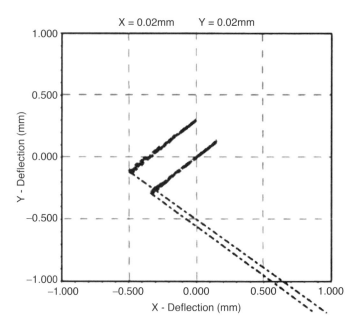

Figure 6.19 Time movements of an individual monitoring point before, during and after the blast.

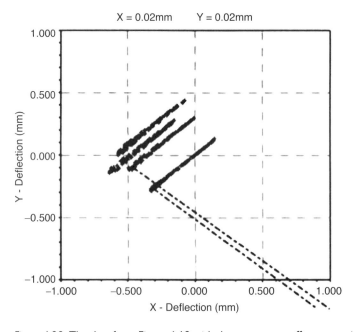

Figure 6.20 The data from Figure 6.19 with the permanent offsets experienced during previous blasts added.

of tunnel deformation, with the accuracy of the deformation data confirmed by precision surveying (Kimmance, 1995). Intervals of logging and their distribution across 24 hours were varied during the trial in order to assess:

a)   tidal influences of the River Thames on the lining;
b)   temperature changes in the tunnel; and
c)   the influence of trains on the stability of the BCS, and any deformations of the lining.

The results are discussed below.

Temperature changes

The thermocouples installed in the system demonstrated that temperature within the tunnel varies with time. A typical 2° Celsius change was observed between operating and engineering hours. Stratification of temperature also occurs, with a difference of 0.5° Celsius between crown and track level being recorded during the early hours of the morning. Examination of displacement variation with temperature over this short period, and also for several months, suggests that for the recorded temperature variation of 7° in the tunnel the system remained stable. This apparent insensitivity to temperature was to be expected considering the relatively small changes in temperature and the use of steel connecting bars (for which the system calibration takes into account temperature-related length changes).

Train transit effects

The location of the trial was such that the trains were just coming out of a bend when passing the system. This in theory places a dynamic vertical and lateral load on the track, and therefore the tunnel lining, and should result in some deformation. Using a logging interval of 30 s it was possible to obtain readings at various stages of a train transit of the BCS location. Data shows movement was detected as a slight (almost negligible) temporary elastic displacement of the lining of up to 0.3 mm in the knee, sidewall and crown of the tunnel. The change in shape of the tunnel is consistent with a primarily vertical load pulling the sides of the tunnel in by depressing the invert, and thereby effectively increasing the vertical dimension. A lateral load due to the curve in the track and braking of the train on this curve offsets this observed vertical extension by 10–15° from the vertical, i.e. the long axis of the deformation is parallel to the resultant of the forces.

   The system had the ability to detect the passage of a train requiring approximately 15–20 s to pass through the monitoring section. In a timed trial period, maximum displacement was shown to occur at mid-train with a rapid rebound to stability once the train passed. The rebound back to initial readings from those when detecting a train demonstrates the stability of the BCS against stray electromagnetic pulses and mechanical vibration.

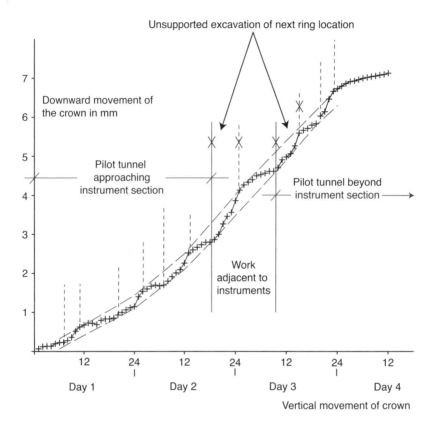

Figure 6.21 Key locations of deformation in a tunnel approaching failure.

The influence of the excavation of a new adjacent tunnel

Figure 6.21 shows an output from the BCS in an existing tunnel where work on a step plate junction tunnel was approaching. It clearly shows the complex relationship between excavation and ring erection in the adjacent tunnel as the work approached, and the move towards stability after construction work had passed. The stabilisation is very rapid.

Figure 6.22 shows the possibility of monitoring the situation as rate-dependent data. Data must be taken at frequent intervals during the construction sequence. The data show rates of movement of the key crown monitoring point in the existing tunnel. It can be seen that the vertical rate increased in steps over the first two days, reaching some 0.42 mm/h by the end of day two. Comparing the tower-like steps with the construction programme clearly demonstrated that the large rates of movement occurred during open excavation for a ring, and fell as the ring was installed. The question at the end of day two was whether this rate would continue to accelerate? In fact, it proved to fall away during days three and four. The data illustrate that great care must be taken with the timing of readings and subsequent interpretation, as individual readings once per hour or every two hours would have given no sense of the problem.

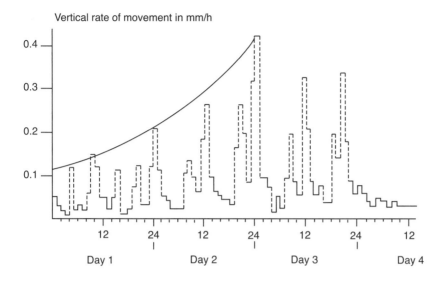

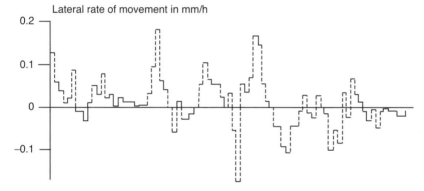

Figure 6.22 The rate of crown movement with time.

Conclusions

The Bassett convergence system of linked beams with mounted electrolevels installed to monitor the cross-sectional deformation of a London Underground running tunnel operated very successfully over a period of 13 months. This assertion is based on the following criteria:

- The system was sufficiently flexible in its space requirement and shape for installation to proceed in a cramped tunnel with many side wall services and train kinematic envelope clearance requirements of 75 mm.
- The system was reliable over the period, with no discontinuity in data except for minor system improvements.
- The robustness of the system was demonstrated by its ability to withstand disruption during tunnel maintenance work involving rerouting services and track replacement in adjacent sleepers. This resulted in abrupt displacements being

recorded but immediate resumption of stability after the initial impact or displacement. Under train operating conditions it was shown that the system was not adversely influenced by mechanical vibration, temperature fluctuations and electromagnetic emissions.

• The sensitivity of the system to short-term displacement was to 0.001 mm movement with a repeatability of 0.01 mm.

During the first three months of operation it became apparent that, under the dynamic conditions of a train passing, the long basal bar spanning below the track was susceptible to oscillation, falsely indicating displacements of up to 3 mm if this coincided with logging intervals. Removal of the beam is not an option if closure errors are required, as in this case. The alternative of damping the beam was therefore adopted. This successfully dealt with the problem.

The results from the instrumented tunnel suggest that the soil/lining-structure forming the tunnel unit is competent and exhibits a small pseudo-elastic cyclical deformation in response to external tidal influences and internal dynamic train loading. It should be noted that if both Δx and Δz displacements are required, a chain of BCS units can be installed on any alignment in any structure. The units were designed for new tunnel construction; however, the author awaits the opportunity to carry out a trial.

Fibre-optic systems

An additional major new piece of instrumentation that should now be considered is the installation of fibre-optic strain measurement, e.g. within structures such as the Mansion House, London. The Fibre BRAGG Grating (FBG) system uses a light interference pattern at points on a fibre; these are created by laser holograms etched into the fibre at defined intervals, which can then match the reference pins in a settlement beam system to form a defined gauge length. If each FBG has the same light-tuned pattern, then a single fibre-optic cable is required for each measurement point. If each individual FBG is different, then a single cable can monitor many hundreds of defined gauge lengths over a considerable distance, as the broad laser beam sent down the cable will reflect the particular unique frequency of the BRAGG and transmit all other frequencies.

The fibre therefore acts as a horizontal strain gauge between the FBGs (see Figure 6.23); changes in the gauge length are measured by comparing the reflected period of each unique signature signal from the two FBG-etched BRAGGs.

The BRAGG consists of a carefully etched length of fibre-optic cable. The etching provides a unique format rather in the manner of a barcode (see Figure 6.24). The etching will return a light signal which represents a narrow frequency spectrum, see Figure 6.25. Note: An individual BRAGG can also be used as an individual strain gauge if the BRAGG length is fixed (from a to b in Figure 6.24) to the material or the member to be monitored. When the unit is then stretched (a–b′ in Figure 6.24) the etchings change, and consequently the reflected frequency plot shifts, see Figures 6.26 and 6.27.

Analysis of this shift will provide the average linear strain along the length of the BRAGG. Because fibre-optic cable expands and contracts with temperature variation

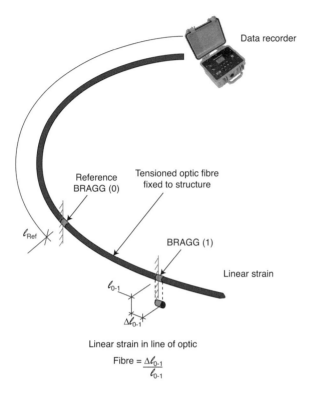

Figure 6.23 Sketch of a fibre-optic system.

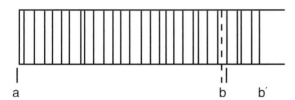

Figure 6.24 Etched pattern on the fibre-optic cable (diagrammatic).

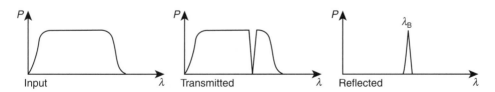

Figure 6.25 Laser input/transmitted and reflected signals at one BRAGG.

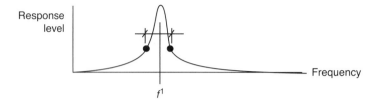

Figure 6.26 Response/frequency curve for a unique BRAGG.

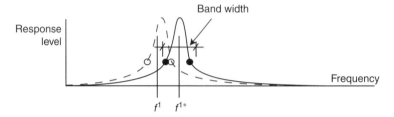

Figure 6.27 Frequency shift on a single BRAGG.

a similar frequency shift will occur, and a free or 'unattached' BRAGG may be used to provide temperature compensation to the measuring BRAGG.

The excitation and analysis unit is a complex system of a controlled coherent light source and sophisticated electronics providing accurate timing of the light travel times and frequency analysis in the optical range, accurately recording the full spectrum of the reflected light. A fast Fourier transform (FFT) is applied to provide the reflected spectra and compare any changes.

It cannot be stressed strongly enough that the data is for a linear strain over the BRAGG length in the fixed direction of the optical fibre. The single unit is simply a strain gauge which can be incorporated in many of the instruments already described to replace the current strain unit, such as the vibrating wire or the linear foil strain gauge. This broadens the use of fibre-optics considerably for use in conventional instruments.

As indicated earlier, if a complete length of fibre-optic cable is firmly fixed to a structural element, preferably pre-stressed, between two BRAGG points, then the travel time difference between the two BRAGGs up to several metres apart can be directly measured for the transmission pulse and the two reflected pulses, and the change in distance between the two BRAGGs can be determined. In this way, it provides the average strain related to the linear displacement from A to B, or the average strain over a considerable gauge length (e.g. 1–5 m, or even greater).

Alternatively, a fibre-optic cable has a natural manufacturing roughness which in itself will provide a reflected light spectra – any short length of fibre having an almost unique reflected frequency spectrum. If the return timing is set for a small fixed period t_1 to $t_2 = \Delta t$, then at a particular distance along the cable the response spectrum for a small length can be examined, the whole cable is examined in Δt increments starting at t_0 to the limit of transmission and reflection.

Then, after loading, the change in the frequency spectra (colour shift) for each length is examined providing a detailed distribution of linear strain. This is equivalent to an almost infinite series of connected BRAGGs. As before, temperature compensation is critical and a second 'reference' fibre may be required. This system requires considerable software power and analysis and is termed 'time domain reflectometry' (BOTDR).

BRAGG units cannot directly measure change in tilt, and are currently unlikely to supersede MEMS units for this application. If, however, a set of three BRAGGs are fixed at one location at three angles round the circumference of a constant cross-section tube (e.g. a thin-walled pipe or a section of inclinometer casing), then looking at the section where the three gauges are attached will provide three local linear strain changes over the three BRAGG gauge lengths, since installation. Figure 6.28 shows a section with the BRAGG lines 1, 2 and 3. In Figure 6.29 the strains measured are shown as ε_{60}, ε_{180} and ε_{270}; all are compressive but different (drawn to scale, inward from the outer perimeter).

$\varepsilon_{\mathrm{mean}}$ is the average compressive strain.

$(\varepsilon_{60} + \varepsilon_{180} + \varepsilon_{270})\frac{1}{3}\varepsilon_{\mathrm{comp}}$ suggests that the tube has experienced some general compression. By removing $\varepsilon_{\mathrm{mean}}$ a residual bending strain pattern remains as indicated in Figures 6.30 and 6.31.

It must now be assumed that the tube bends in a perfectly elastic manner and that plane sections remain plane. We need therefore first to determine the axis of bending. The strains are plotted across the 180° axis. They have a very curved profile, see Figure 6.31(b) (non-linear bending strains).

If the section is strained in simple bending the three strains must lie on a straight line through the centre O. The axis has been rotated in Figure 6.31(c) and the strains replotted onto this axis A–A, and the strains form a straight line as required. Axis A–A

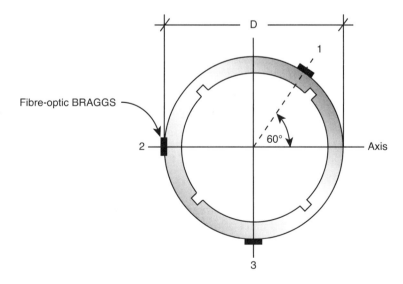

Figure 6.28 Fibre-optic cables located on an inclinometer casing.

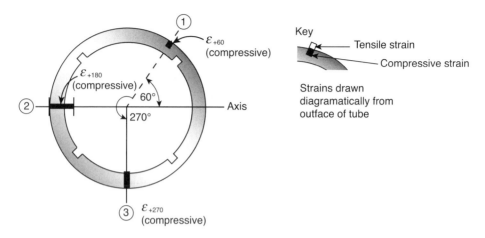

Figure 6.29 Strain measured in the three BRAGGs.

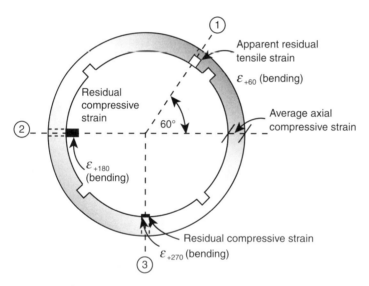

Figure 6.30 Diagrammatic representation of the bending strains round the tube (drawn inward (compression) from the inner perimeter).

is then the axis of bending and the maximum tensile and compressive strains on the section circumference will be strains $\pm\varepsilon_{max}$.

The axis of bending has now been identified and the maximum strains determined across the section. They can be used to define a curvature based on the length of the BRAGG systems gauge length ℓ_b, the diameter of the tube D and the maximum strains assessed above. This is shown diagrammatically in Figure 6.32.

For example, if the inclinometer casing used is 70 mm in diameter and over the location of the BRAGG the maximum bending strain is 0.001 (i.e. 1/10 of 1%), the radius of bending is 35 m.

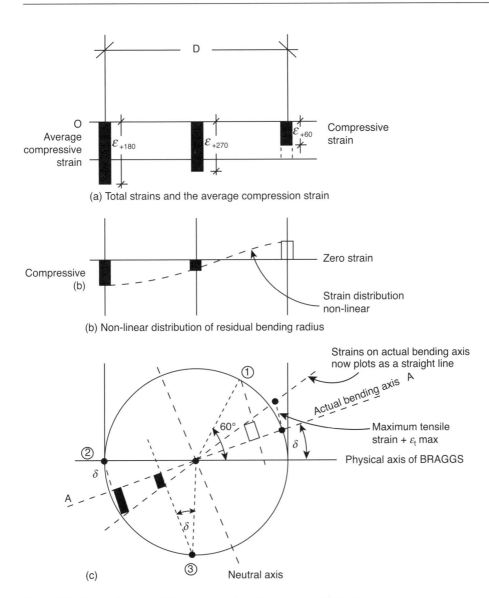

ε_{+180} ε_{+270} ε_{+60}

(a) Total strains and the average compression strain

(b) Non-linear distribution of residual bending radius

Figure 6.31 Determination of the bending axis by linear strain distribution.

There is a serious limitation to the system, in that there is no initial absolute orientation from which to start the lateral displacements from the curvature components. Ideally the tube or section used to mount the BRAGG sets must contain no joints as at (even perfect) joints sharp changes in orientation can occur. In-place inclinometer systems bridge these joints and always refer back to a gravity reference. So for a fibre-optic arrangement using 3 m casing lengths, the curvature principle will enable the curvatures from three BRAGG sets at 1 m to 2 m to determine the bent shape from joint N to N + 2; however, the tangential angle cannot be guaranteed across the joint at N + 2, the lateral displacements being lost; see Figure 6.33.

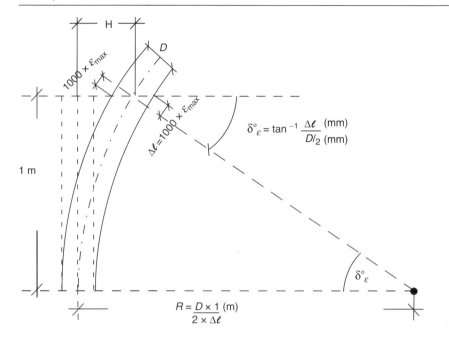

$$\delta^{\circ}{}_{\varepsilon} = \tan^{-1} \frac{\Delta\ell \ (mm)}{D/2 \ (mm)}$$

$$R = \frac{D \times 1}{2 \times \Delta\ell} \ (m)$$

Figure 6.32 Diagrammatic relationship between measured maximum strain E_{max} and R.

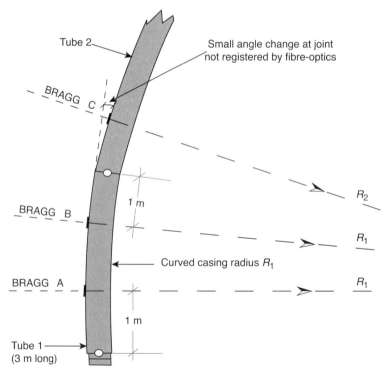

Figure 6.33 Diagrammatic illustration of unmeasured angle changes using fibre-optic strains.

Chapter 7

Instrument systems for tunnelling projects

This chapter aims to provide an example of the overall interaction between theoretical analysis, soil characteristics, predicted behaviour and measured field behaviour using a variety of instruments.

Initial assessment (experience)

Tunnelling, and in particular urban tunnelling, is becoming a major component of modern infrastructure improvements. In the last three decades there have been major advances in tunnelling techniques, particularly in the development of many new forms of tunnelling machines (e.g. earth pressure balance machines, multi-headed rotary cutters, very large shields with built in excavators) and the development of sprayed concrete, together with NATM (New Austrian Tunnelling Technique), instrumentation and support techniques.

Rapid mechanical excavation and installation of lining rings, and the subsequent grouting to seal the tunnel all limit the 'face loss' and the associated surface settlement. These, together with compensation and compaction grouting, have changed the situation, and require a considerable increase in installation of instrumentation to provide an understanding of the new conditions that result from these new and innovative techniques.

Urban tunnelling in particular is critical, as it influences many third parties owning properties and facilities on and within the ground above and to the sides of the new tunnel work. Deep basements and deep piles can be significantly affected, not just by new tunnelling, but when they form part of a new development adjacent to or above an existing tunnel.

Prior to new work, an assessment will be made using empirical and numerical methods. Commonly only two-dimensional assessments are made, but in complex or critical conditions three dimensional-finite element (FE) analysis (e.g. ABACUS) is used. The outcomes of these studies are predictions of the stresses, stress changes and displacements throughout the soil mass and the tunnel linings at different stages and times during construction.

There are major limitations to these predictions, mainly because the soil details and characteristics are often severely restricted (e.g. linear elastic, perfectly plastic), and the dissipation of pore pressure with time and the subsequent consolidation/settlement and lining stress changes are difficult to assess. They do however provide guidance on the likely critical locations.

Field instruments provide a real check on these predictions and can help improve future assessments. The numerical work can indicate where field instruments may be of most use and the level of accuracy required.

The author co-authored a paper presented to the Institution of Civil Engineers in 1999 and presented a general paper on these topics to 'Tunnelling 2007': updated versions of these are included in this chapter.

Tunnelling and instrumentation

There are now many records within the literature of measured stresses and displacements in and around tunnels. The author was involved in the data recording and assessment of: a) The settlement of the Mansion House, London (Forbes *et al.*, 1994), the implications of which were discussed within the engineering context by Powderham (1998); and b) the trial NATM tunnel at Heathrow Airport, London (Dean and Bassett, 1995), also documented by New *et al.* (2000), being typical of some of the fuller instrumentation exercises reported. The author will try to set out the full philosophy of planning, installing and assessing an instrumented tunnelling project, hoping that the reader can use it to assess their own priorities and clearly appreciate the confidence it can provide, and what they can learn and carry forward to future projects.

Preliminary predictions

Any project for which field instrumentation is proposed needs a class A prediction and some careful thought needs to be given, using the database of prior experience, to identifying the sensitive areas. The real problems that arise in the field are soil/structure interaction problems where an affected structure is not within the client's ownership or control. It is only rarely a safety problem. The observational method proposed by Powderham (1998) for controlling construction by instrumentation involves the concept of risk judgment, risk is most commonly judged against a serviceability assessment. Settlement of structures due to tunnelling is normally the largest of the deformations, it is not however the only form of deformation that could prove critical.

Assessing deformation and the resulting displacements is the most difficult prediction problem in the non-homogeneous, non-isotropic, three-dimensional space that is a real soil. Adjusting deformations due to the responding influence of the structure adds an order of magnitude to the difficulty.

The start point for preliminary assessment is inevitably the accumulated experience developed for a green-field site as set out by Attwell and Farmer (1974) and summarised in a most useful form by O'Reilly and New in their TRRL documents in 1982. The idealised settlement trough is the progressing depression shown in Figures 7.1 and 7.2. The active zone is formed by a three-dimensional conic travelling with the tunnel face and finally leaving a two-dimensional trough in its wake. The settlement of this two-dimensional trough is assumed to be in the form of an error function curve:

$$S = S_{max} \exp S = S_{max} \exp \left(-\frac{y^2}{2i^2} \right)$$

shown by O'Reilly and New as Figure 7.3, and where y is the distance sideways from the centreline.

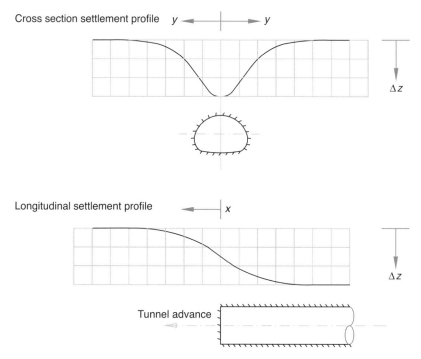

Figure 7.1 Sections through an idealised settlement trough.

The key variables being: the depth of the tunnel below ground; the 'volume loss' resulting from the tunnelling technique, the soil/water regime and a trough width parameter i, which is dependant on soil type, where $i = kz_0$ (z_0 being the depth to the centreline of the tunnel and k a soil constant).

Figure 7.4 shows model data obtained in centrifuge tests at Cambridge University, UK, clearly differentiating clay and sand behaviour. The value of k is again based on experience. Two key figures, again from O'Reilly and New, slightly simplified by the author (Figures 7.5 and 7.6), provide i approximately $0.5\ z$ for tunnels in clays up to 20 m deep and up to 0.3 m in sands. If settlement displacement vectors are aligned on a cross-section of the tunnel then an assessment can also be made of the horizontal displacement and consequently of surface strains. In particular, attention is drawn to the combined hogging and tensile strain that occurs between $1.5i$ and $2.0i$. In clays it is assumed that the volume of a section of the settlement trough is the same as the extra volume excavated in the tunnel: 'volume loss'. O'Reilly and New also provide the data in Table 7.1 – typical data ranges for the 'volume loss' experienced in typical tunnel monitoring.

If the concept of $k = $ constant is extended to any depth below the ground surface it would result in a simple conic form, data such as in Figure 7.4 show that this does not represent the real behaviour.

Model research data and field data from several sources clearly prove that the conic assumption is far from the truth. Adopting this simple analogue for deep basement buildings, services or piles may be unacceptably inaccurate. The most useful approach

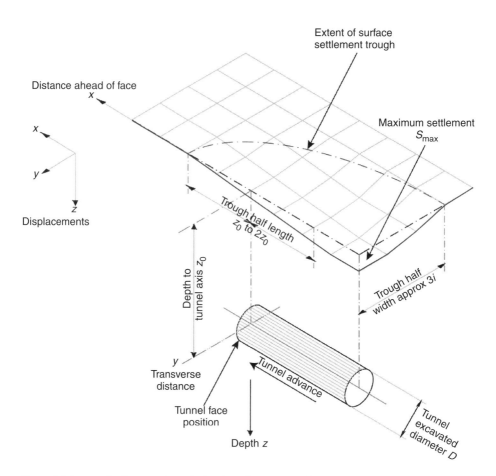

Figure 7.2 Isometric diagram of an idealised settlement trough.

to foundations and services that are embedded in the soil above a tunnel is to modify the value of k for the appropriate horizon level. A useful diagram (Figure 7.7) for this adjustment is provided by Mair *et al.* (1993) where z is the depth of the specific horizon from ground level and z_0 is the tunnel depth to the centreline.

Returning to the surface settlement but now considering the influence of horizontal strain, the classic work remains Boscardin and Cording (1987), following a key paper on settlement by Burland and Wroth (1975). Three figures from Boscardin and Cording have been reproduced as Figures 7.8–7.10 and it is hoped that they are self-explanatory.

All the above comments are related to a simple single tunnel construction. Critical problems are commonly associated with complex situations, involving running and station tunnels, cross passages, even vertical and inclined shafts, and in particular differing tunnelling methods. Dividing the tunnels into component blocks and assuming a specific volume loss can be associated with each component, and distributed to any horizon or surface as a circular conic (Figures 7.11 and 7.12), or corrected for level according to Figure 7.7. The volume loss being based on a radially symmetric

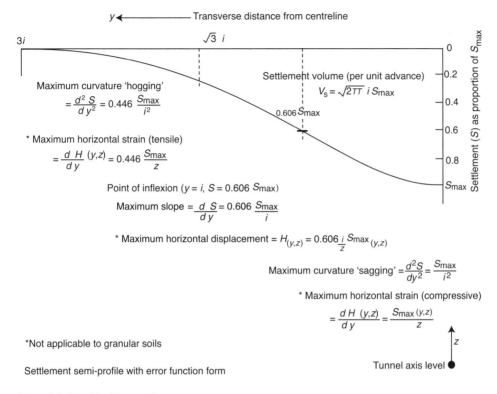

Figure 7.3 Simplified inverted error curve.

corrected error (Figure 7.13) curve

$$S = S_{\max} \exp\left(-\frac{r^2}{2i^2}\right) = \frac{\text{volume loss}}{2\,\pi\,i^2} \exp\left(-\frac{r^2}{2i^2}\right)$$

where r is a local dimension for an individual excavated volume with respect to a surface grid which is associated with the tunnel axis. The grid points are locations at which settlement of each component are superimposed, the relevant i value, and the level of any required horizon being fixed for each zone of excavation. The tunnel section can be advanced in steps of l grid spacing, and all the individual components can be numerically summed on the basis of super position. The author is aware that consultants such as Mott MacDonald have such an approach in-house.

The horizontal strains can be deduced on a similar basis if the displacement vectors are all assumed to align to the centroid of each component of the excavated volume (Figure 7.13). The numerical summation is not the easiest of the techniques, but working on a reasonable grid, the system is successful. It allows better predictions for large or shallow tunnels, and even shafts, sloped tunnels and excavations where the diameter changes and construction methods can all be assessed. For instance, if the elements shown in Figure 7.11 are dimensioned to all have identical volumes (as shown in Figure 7.14), and different volume losses are ascribed to represent for instance the

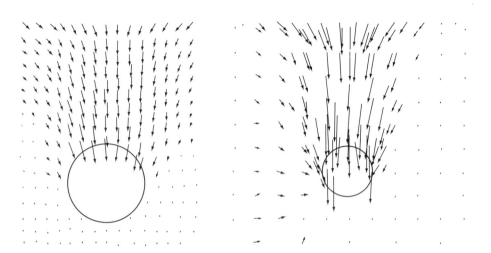

Figure 7.4 Displacement vectors in the vicinity of tunnel construction.

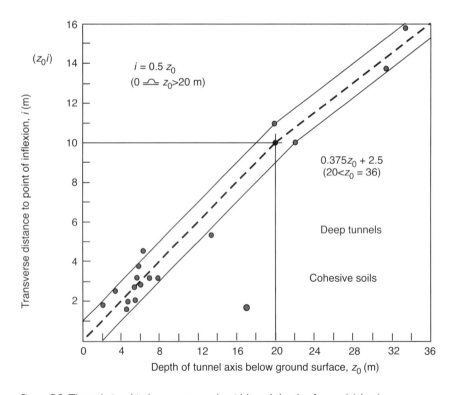

Figure 7.5 The relationship between trench width and depth of tunnel (clays).

Table 7.1 Data ranges for the 'volume loss' typically experienced in tunnel monitoring.

Ground conditions	Ground support method in tunnel	Trough width parameter constant, k	Ground loss $\frac{V_s}{V_{exc}}$ %	Remarks
Stiff fissured clay	Shield or more	0.4–0.5	½–3	Considerable data available; losses normally 1–2%
Glacial deposits	Shield in free air	0.5–0.6	2–2½	
	Shield in compressed air		1–1¼[a]	Compressed air used to control ground movements
Recent silty clay deposits	Shield in free air	0.6–0.7	30–45[b]	Failure or near-failure conditions usual
$cu = 10$–$40\,kN/m^2$	Shield in compressed air		5–20[a]	Some partial face values included

[a] Compressed air is less common now than 20 years ago, being again replaced by an EPB machine.
[b] The 30–45% values are untenable and 'free air' would no longer be employed, being replaced by an earth pressure balance (EPB) machine.

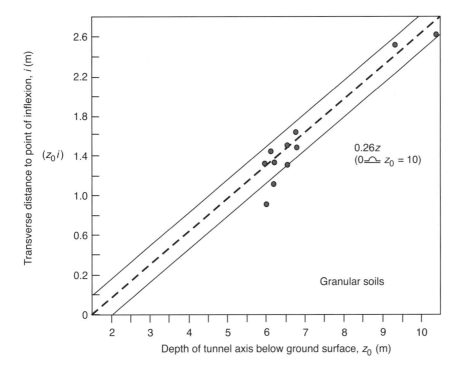

Figure 7.6 The relationship between trench width and depth of tunnel (sands).

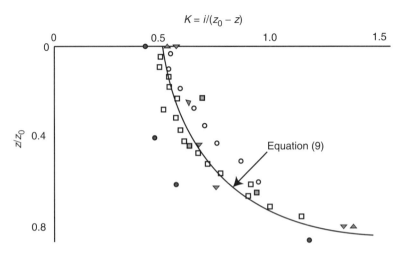

Figure 7.7 Variation of *i* below ground level for adjusting the settlement trough with depth.

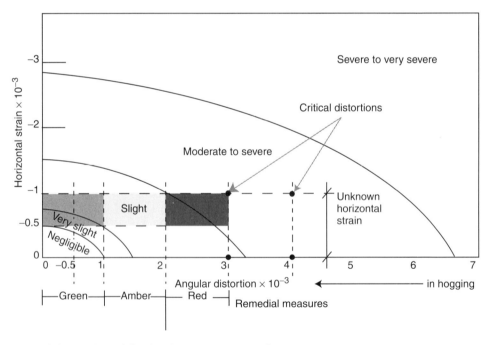

Figure 7.8 Boscardin and Cording damage assessment diagram.

'type C' full face NATM construction used at the Heathrow trial tunnel, individual volume losses can be given to each element, as shown in the tabular part of the figure.

The result of the surface angular distortion and the horizontal strains calculated can be applied to the structure as if it is a semi-stiff beam attached to the soil surface; any surface strain alters the location of the overall neutral axis of the 'beam structure'.

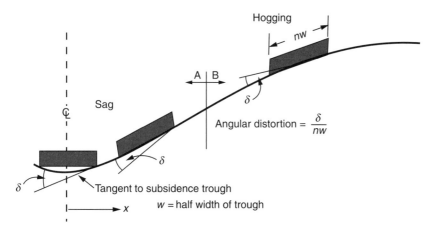

Figure 7.9 Trough distortion.

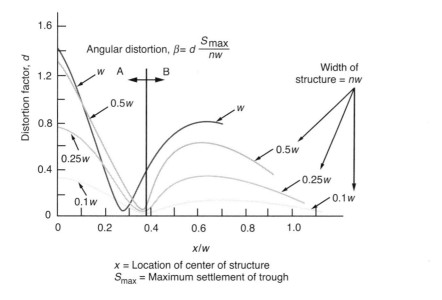

Figure 7.10 Angular distortion versus horizontal location.

This again is a conservative assumption as the real soil/structure interaction will modify the green-field deformations. In particular, the load distribution on the foundations may well change, the outer zones falling, the central area increasing and the resulting angular distortion being flattened. If the foundations are linearly reinforced, there may well be relative slip between the soil and the foundation. As far as the author is aware there are currently no simple rules to deal with such structural modifications. The incorporation of an elastic frame or beam system to represent the structure, and slip elements to mimic the foundation/soil interface in a numerical program, can go some way towards providing a better technical answer as outlined in the next section.

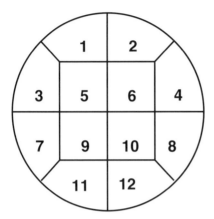

Figure 7.11 Division of a tunnel into sectional blocks.

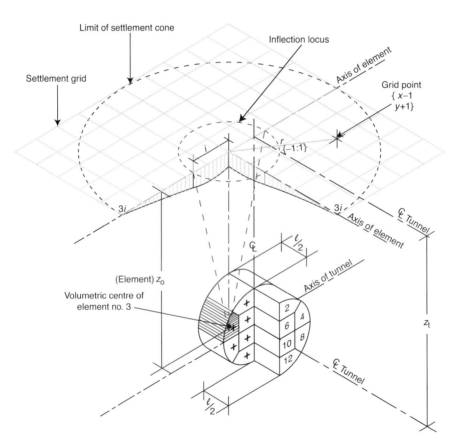

Figure 7.12 Settlement conic for element 3.

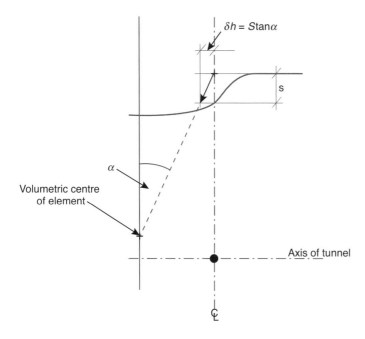

Figure 7.13 Assessment of component of horizontal strain.

Numerical assessment

Having made preliminary predictions based on this well-documented green-field experience, more-sophisticated predictions can be obtained by numerical methods. Both finite element and finite difference codes exist, many consultants even having their own in-house programs. Full three-dimensional programs are very complex to set up in order to develop a progressive tunnelling technique (their run times are long and they are still limited by the constitutive soil model that is incorporated), but ABACUS can do this. Two-dimensional simplification is commonly adopted on the basis that, if a two-dimensional simplification shows no problems, then the situation is probably satisfactory. The author has had experience with FLAC, ABACUS and CRISP, each having their individual difficulties and limitations. The progressive tunnelling is represented by the softening and eventual removal of the soil elements within the tunnel envelope and the insertion of a lining; its progressively increasing stiffness or subsequent grouting being modelled. Time is always a critical factor as there could be progressively increasing volume loss with unsupported time, followed by the insertion of the tunnel lining, which itself may progressively stiffen (e.g. NATM). These time-dependent problems are dominated by the development and time dissipation of negative pore water pressure, and the separation of effective stress and pore water pressure is a key component of the soil model.

The assumed soil models describing the effective stress response can also vary greatly, from the most simple linear elastic, through linear elastic/perfectly plastic, associated or non-associated Mohr Coulomb response, to a Roscoe and Burland cam clay formulation, and (in the author's current research) a 3SKH formulation where the cam clay

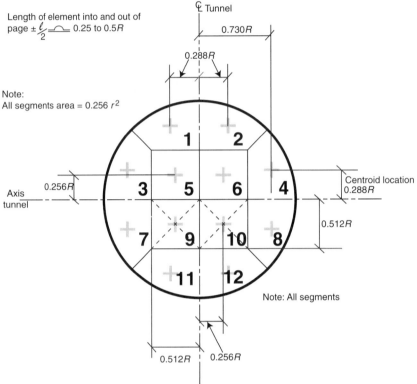

Example for a NATM excavation

	Crown	Left side	Right side	Centre	Invert
Excavation sequence	1 & 2	3 7	4 8	5. 6. 9. 10	11 12
Individual volume loss %	3.0	2.5	2.5	0.50	1.5*
Actual volume loss % based on identified area	6.0	5.0	5.0	2.0	3.0

* This value can be critical and will increase considerably if the invert is not closed rapidly after blocks 5, 6, 9 and 10.

Total

$$\left.\begin{array}{r}6.0 \\ 10.0 \\ 2.0 \\ 3.0\end{array}\right\} = \frac{21.0}{12} = 1.75 \text{ average volume loss}$$

Figure 7.14 Equal area elements.

state boundary surface has an appropriate (considerably modified) plastic potential, supplemented by a nested but moving elastic surface and history surface.

However sophisticated the numerical model may be, fundamental input parameters are still chosen from experience. The *in situ* stress conditions $\sigma_h^l = K_0\sigma_v^l$ are often not physically measured during the site investigation, although equipment such as the Cambridge self-boring pressuremeter can provide this value (usually between 1.5 and 2.5 for London clay), although the author has seen many numerical calculations based on $K_0 = 1$. The plastic limit will usually be assumed as C_u and the Mohr Coulomb criteria as ϕ^l; however, the E value and v (Poisson's ratio) value are not so easy to calculate. Again, many consultants have in-house databases relating E_u for London clay to C_u, and v is commonly assumed to be 0.3 to 0.35 for drained conditions (putting $v = 0.5$ in an undrained analysis appears to cause inordinate numerical trouble). An effective stress calculation may require the water to be given a reduced bulk modulus as a full $B = 1.0$ response does not appear to actually take place in real life. A careful stage-by-stage development of the stress and displacements in the soil mesh has to be carried out to achieve a representation of the *in situ* condition before the tunnelling is even modelled. Any structures must have their overall stiffness inserted, and interface elements need to be incorporated where appropriate. (It should be noted that there are often limitations to the detailed location of interface elements.) The modelling still requires a decision on volume loss and how to apply it to the tunnel wall; construction timing and finally the installation of the lining both need to be representative of the actual process. The structure is almost always assumed to be a complete homogeneous ring; however, the ring stiffness must be reduced if there are joints, the historic Curtis–Wood reduction remains relevant; $I_{\text{apparent ring}} = I_{\text{solid ring}}.\ (4/N)^2$ where N is the number of segmented units, Muir-Wood (1975).

The author would recommend that, before running any computer program, a qualitative distortion pattern is sketched and compared with the form of the computer output. The author has more than once seen discrepancies that invariably prove to be due to lack of careful consideration of the initial setting up of the computer program's *in situ* assumptions. Figures 7.15–7.19 show a typical CRISP assessment for the complex concourse tunnel at London Bridge Station, London, carried out by the author's colleague Professor A. Swain, with the various useful outputs shown.

The key variable in most of this numerical work is the ratio of the shear modulus G_{struct} of the tunnel lining to the G_{soil} of the soil. It is recommended that a parametric check should always be carried out, increasing the $G_{\text{soil}} \times 5$ and reducing it to $G_{\text{soil}}/5$ to check sensitivity. Also, checks should be made using the best and the worst possible volume-loss values, e.g. 0.5–1% for the best and 1.5–2.5% or even 3% for the worst in London clay.

Figures 7.20 and 7.21 show some of the numerical data obtained by PhD student Y. J. Lee (2004) for tunnelling in granular material adjacent to an *in situ* loaded diaphragm wall. In this study, excessive tunnelling volume loss was used to investigate loss in both the side-friction load and the end-bearing load of the wall associated with settlement around the tunnel, and the potential failure mechanism that would develop. The numerical data was assessed against a physical model, and failure against both an upper- and a lower-bound limit analysis.

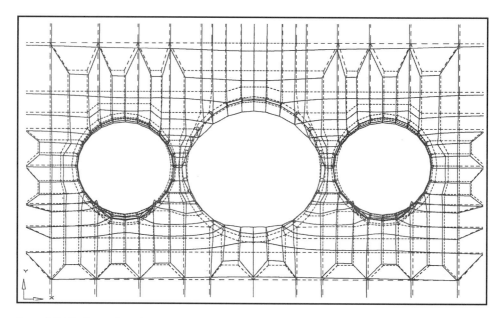

Figure 7.15 Deformed mesh.

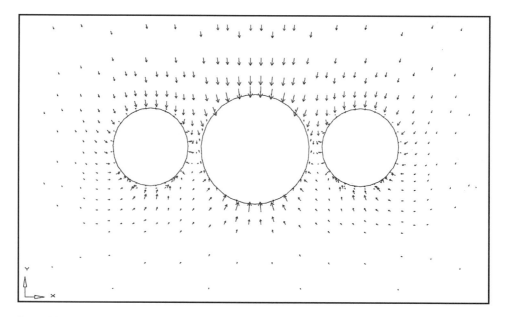

Figure 7.16 Displacement vectors.

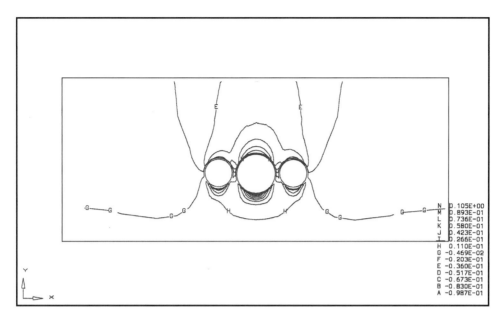

Figure 7.17 Vertical displacements.

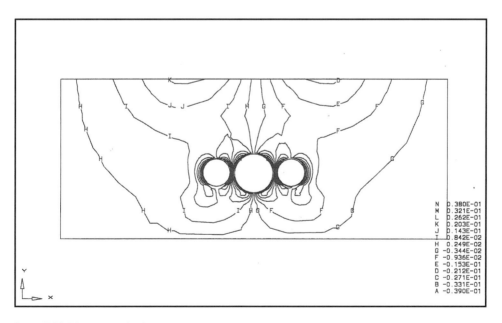

Figure 7.18 Horizontal displacements.

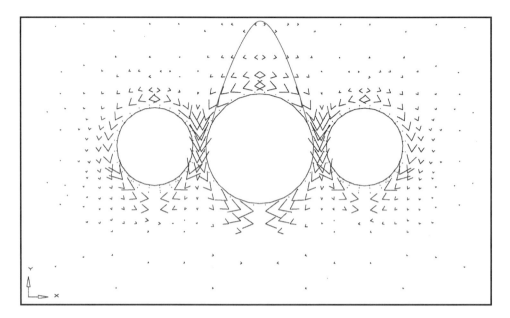

Figure 7.19 Zero extension directions (potential failure planes).

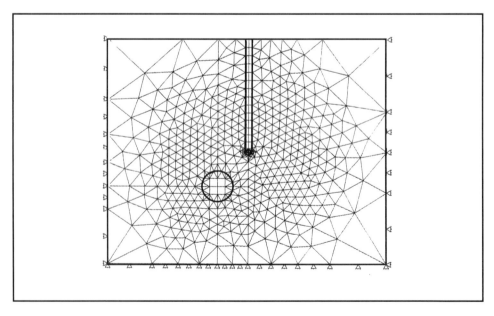

Figure 7.20 Finite element mesh for a wall/tunnel interaction problem.

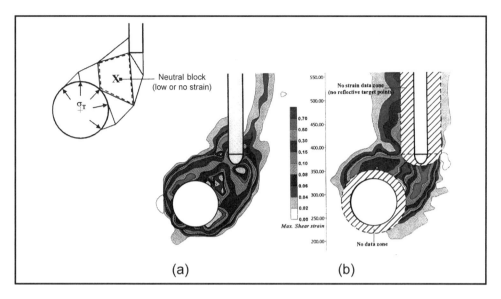

Figure 7.21 Shear strain data for a pile/tunnel interaction: (a) finite element analysis, (b) physical model.

At the end of the prediction and project assessment process, the most useful outputs for the planner are:

a) The complete displacement field between the tunnel and any structure or existing service requiring damage assessment.
b) The vertical displacements, the horizontal strains resulting in changes in the bending moment and column loads within these structures.
c) The development and time dissipation of pore pressures within the soil.
d) The deformation of the tunnel lining and lining stresses with time from as early in the construction process as possible.

Instrumentation

The above outputs constitute a class A prediction. They should indicate areas that show the greatest displacement, strains or stress changes and enable a decision that:

a) The predicted behaviour is acceptable and no significant problems or damage are likely. Instrumentation may be used to confirm this, to provide experience and to improve future predictions.
b) The response is close to potential problems. Instrumentation, together with the 'observational method', with planned intervention available when necessary, is recommended.
c) The response is unacceptable. Specialist real-time instrumentation and compensation measures must be included in the work.

Situation c) may cover many specialist operations, from compaction grouting, ground freezing, compensation grouting, to even underpinning prior to tunnelling, the influence of which will also need to be numerically assessed and key locations identified and specialist instrumentation planned.

Situation b) represents an interesting problem as the feedback from the instrumentation will provide the key to a learning sequence. The flexibility of the approaches and significant savings in terms of both costs and time have been well set out and established by Powderham (2002).

During progressive observation, the data must be monitored at specific, regular time intervals, particularly during each critical stage of enhanced-risk construction trials. These may be as frequent as every two minutes and rarely less than two a day. A final working sequence must be agreed between the parties and the key instrument control values identified. The data must be assessed in real or near real time by an experienced engineer. Furthermore, the numerical methods should be updated on a 'class C' prediction basis to adjust the various material parameters to match the progressing instrumentation data. The learning trials should provide a real set of data changes that are perceived as acceptably safe (green), a set of alarm thresholds for which intervention must be set in place (orange), and when urgent intervention must be carried out (red). All future work is then judged against these agreed data changes.

Situation c) divides into two forms: i) serious remedial action prior to any tunnelling, e.g. compaction grouting, underpinning, etc. This too will require the re-use of the numerical program and modifications of the relevant properties to prove the remedial work returns the problem to situation a); ii) Compensation activity during the actual tunnelling operation. This probably requires an even greater level of instrumentation measuring quality and speed of response than the b) situation. It must be based on units capable of recording small rates of change of deformation or stresses, and an automatic system that warns when any absolute magnitude or rate is exceeded, as compensation work involves potential safety in both the tunnel and the deforming structure. An engineer must be responsible for making on the spot decisions in what can easily become a 'catch 22' situation.

Finally, the type a) situation suggests an acceptable position and that all parties are confident no problems will arise. Is instrumentation necessary? In terms of the individual project, possibly nothing more than levelling surveys at suitable intervals (e.g. at the beginning, middle and end of construction, and six months later) could be all that is required. Enlightened consultants realise that this is short-sighted. Many quite unconfirmed parameters are assumed in the preliminary assessment and in any computer program. Basic instrumentation within or near the tunnel and in the affected building could enable key values such as volume loss, the value of i and any time dependencies to be reassessed. Useful real data can be obtained on the modification to the green-field site settlement profile due to the building stiffness, and can be back-analysed on the basis of the instrumentation data. Such data could consist of some of the following:

i)   manual inclinometer and settlement measurements in the ground close to the tunnel;
ii) convergence within the tunnel;

iii) a ground and building settlement profile; and
iv) pore pressure measurement round the tunnel.

These would be taken:

1) before any work,
2) once a day during the passage of the settlement trough, and
3) some time after completion, together with an accurate construction log.

These types of data are obtained with conventional surveying and hand-read classic field instrumentation equipment. Records can be assessed and experience enhanced within the office after completion of the project.

Field instrumentation layouts (ideal)

Situations b) and c), above, are very different from situation a). It is almost certain that any structure in these two categories will really need both detailed settlement (to provide actual distortion) and linear foundation strain to be recorded. Many b) and c) situations may benefit from stress change measurements at the base of any load-bearing columns. All readings need to be recorded, processed, presented and assessed in real time. Data therefore needs to be datalogged and processed on site, and should be assessed immediately. It is now possible, with web-based computer systems, for the data to be presented to all concerned parties at any time in graphical format, such as a settlement-against-location plan for a specific time, and a specific item of data for a key point against time.

All data from structures must therefore be digital. Until fairly recently, this required quite significant power and extensive wiring. This is the key element that has rapidly changed in the last five years; many instruments can now be powered and be read by stand-alone short-range radio transceivers forming part of a mesh network, or now even by secure commercial WiFi systems, getting rid of the easily damaged forest of wires which used to cover instrumented sites.

Convergence

The excavation of a tunnel involves the advancing of an open soil face, sometimes supported by pressure provided by a drilling mud or more commonly by pressurised soil slurry (these days rarely air pressure). The wear on the cutting/excavating equipment can be excessive and replacement techniques have to be adopted. This can temporarily remove the support pressure. Furthermore, the lining system has to be applied (e.g. sprayed concrete) or erected (e.g. rings). Both of which leave the perimeter of the soil unsupported or only partially supported for some period of time, until the sprayed concrete hardens sufficiently or the ring system is grouted.

During this period (and in fact for some time afterwards), while stress redistribution takes place, the walls of the tunnel will move inward. If the lining support is not theoretically ideal either in time or in uniform stiffness, or due to later remedial work, this inward movement will not be symmetric.

The non-symmetric deformation of curved surfaces can lead to dangerous situations, particularly in partially completed NATM sequences. These circumstances have long been recognised and the use of cross-tunnel measurements has been a key to checking such situations.

The tape extensometer shown in Figure 7.22 is an accurate (to 0.1 mm with operator care and experience) tool that is used to measure across a tunnel from reference hooks set in the tunnel lining, a typical array is shown in Figure 7.23.

All the distances in Figure 7.23 from A and G are measured with a check B–F. This is in effect a triangulation system allowing all (Δx and $\Delta y)_n$ displacements to be monitored. Points A and G are usually assumed to be stable, as most displacements are between B, D and F. However, the method is labour intensive and reaching points C, D and E usually requires either access equipment or specialist tape location tools. The key time for measurement is immediately after the lining is just completed and ongoing tunnelling work is just ahead of the measured section. Conflicting interests between tunnellers and instrumentation engineers can cause friction. If an incipient failure is developing, then *no* manpower should be within the new tunnel works. Under these circumstances an automatic system would be a significant advantage. As a result of the Heathrow Airport tunnel collapse in 1994, the author developed the

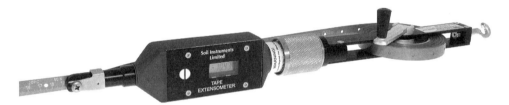

Figure 7.22 A digital tape extensometer.

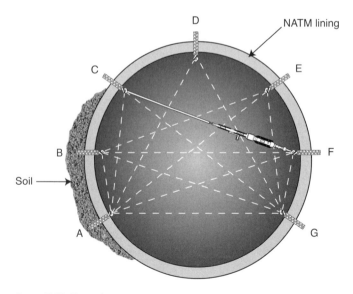

Figure 7.23 Typical tape extensometer monitoring array.

Bassett convergence system (BCS), which has been described in detail in Chapter 6. This originally required hard wiring to an adjacent datalogger, but can now use a mesh network radio and WiFi systems explained in Chapter 8, allowing rapid, minimally intrusive installation.

Structures and deformations

Figure 7.24 shows conventional levelling survey points for the area surrounding the Mansion House, London, this system could now be enhanced by using total station monitoring. Figure 7.24 also shows the local electrolevel beam and tilt array set out in the basement (Forbes *et al.* 1994). This is a typical structural layout measuring settlement. This layout concept remains valid but could now be considerable enhanced for horizontal strain by the inclusion of a complementary fibre-optic system (Chapter 6).

The source of all ground movements and stress changes within the ground and between the tunnel and the surface are due to the tunnel construction itself. Any problem within the tunnel will be the key to all other responses, as the Arup assessment of the hand-measured convergence data after the Heathrow collapse clearly demonstrated. But, as indicated earlier, hand measurements within a tunnel can be very disruptive to the excavation process and, in the Heathrow situation, potentially hazardous to the reading team. The BCS, which can now be rapidly installed and wirelessly connected, measures the full deformed shape of the tunnel (see Chapter 6) and would provide a safer route to obtaining data in a dangerous situation.

The reference pins could be installed into the soil or with any reinforcement rebars prior to the spray concrete. The system could be fitted immediately spraying is complete. But any convergence movements cannot be measured until after the tunnel face has just passed. Absolute measurements of the tunnel deformation can therefore only be assessed from within the soil mass outside the excavated boundary. Conventionally this has been done using a system of inclinometers and settlement gauges as adopted for the Heathrow trial (Figure 7.25). The major change from hand reading such a layout has been the adoption of in-place inclinometer systems which record automatically, complemented by rod settlement units with an automatic reading head, both units now being wirelessly connected. However, in view of the predicted deformation patterns in the near vicinity of a tunnel the author would suggest an alternative layout (Figure 7.26) of automatic inclined in-place inclinometer systems based on MEMS accelerometers, set at 30° or 45° to the vertical.

Global settlement at ground level

Conventional area survey levelling such as those carried out at the Mansion House is shown in Figure 7.27. It is now automated using self-levelling instruments incorporating datalogging and bar coded staffs, or by using total station surveys which will provide lateral displacements as well as settlements. Figures 7.28 and 7.29, show two typical settlement contour plots of the trough approaching the Mansion House. Figure 7.30 shows it pictorially. Comparison with the tunnelling progress reports and the shape of settlement changes can provide a key check on the pre-construction prediction allowing both *i* and face loss to be reassessed.

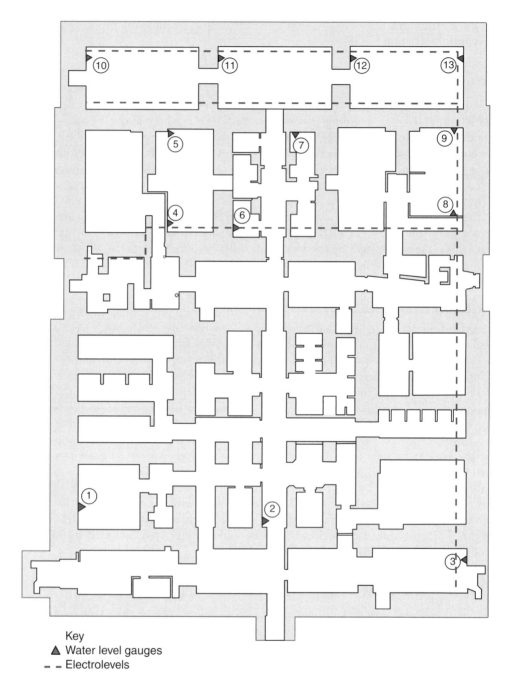

Key
▲ Water level gauges
– – Electrolevels

Figure 7.24 The Mansion House, London, basement.

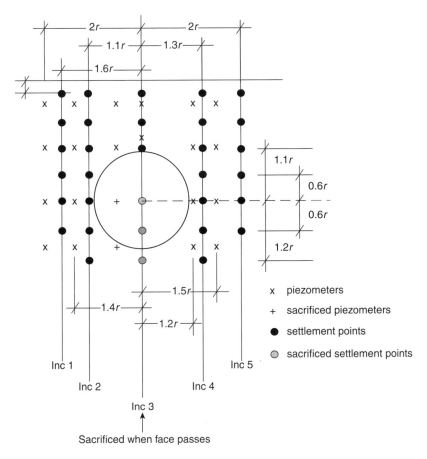

Figure 7.25 Typical layout of inclinometers, settlement gauges and piezometers around a trial tunnel.

Within the structure, the vertical distortion can be recorded by the use of either electrolevel or MEMS accelerometers attached to beams 1.5 to 3 m long, mounted on reference pins on the floor or preferably on basement walls. The Mansion House layout shown in Figure 7.24 remains a good guide; side, front and cross walls all being instrumented as the building plan was at an angle to the tunnel alignment. At the Mansion House these chains of accumulated data were backed up by absolute settlement values recorded by a few critically sited water settlement cells and the resulting settlement contours were plotted as in Figure 7.31, indicating the distortion of the structure. The spacing of the 1 mm interval contours clearly indicates the critical area of angular distortion (shown shaded). It should be noted that water level data are absolute with reference to the remote datum but involve considerable plumbing. The beam system is now capable of wireless interrogation; the water level system has not been updated as the wires can be routed with the plumbing.

Installing fibre-optic systems can provide accurate and detailed strain measurement, which can resolve the detailed deformation of stiff steel structures. It must however

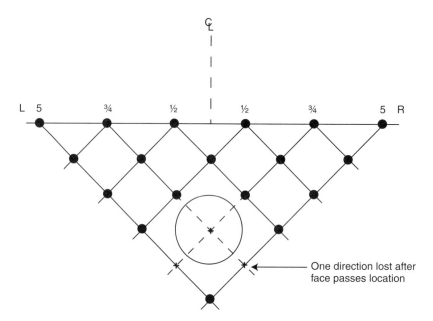

Figure 7.26 Potential 45° inclined inclinometer layout.

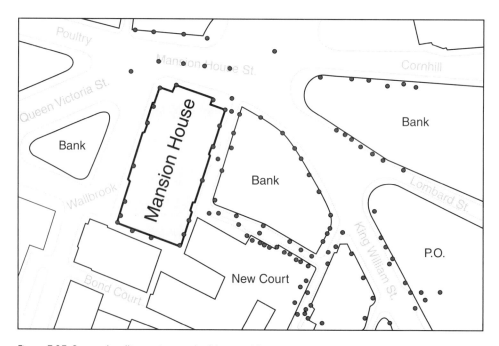

Figure 7.27 Survey levelling points at the Mansion House.

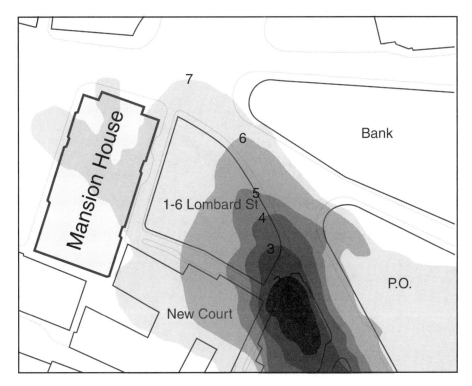

Figure 7.28 Developing settlement trough at the Mansion House, approaching.

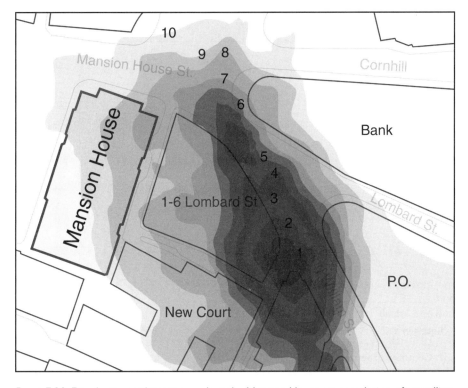

Figure 7.29 Developing settlement trough at the Mansion House, at completion of tunnelling.

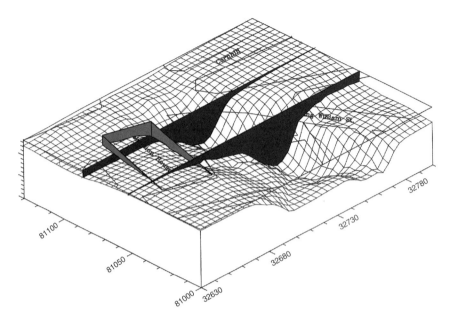

Figure 7.30 Mansion House area global settlement as a three-dimensional plot.

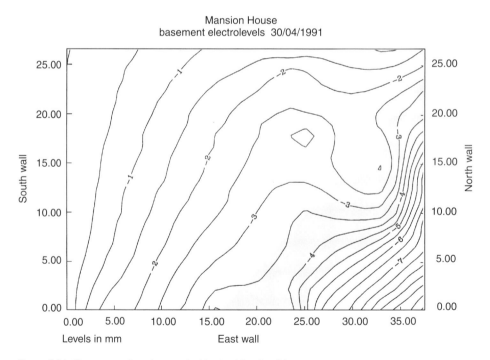

Figure 7.31 Contours of settlement inside the Mansion House.

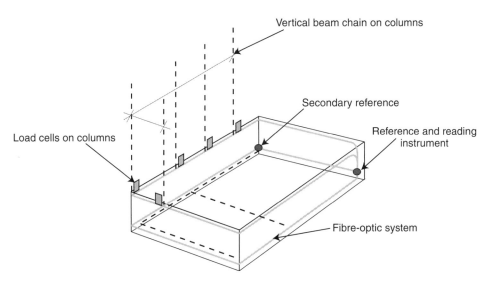

Figure 7.32 Idealised layout of part of a sensitive structure.

be stressed that both beam and fibre-optic systems give one-direction strain measurement at one place. If two lines of strain are measured on a homogeneous section
at a known parallel separation the optical systems can be used directly to calculate
combined compression or tension and curvature, as required for a Boscardin and Cording check. Figure 7.32 shows the possible layout for a fully instrumented basement
structure.

In very tall or particularly sensitive structures, lateral displacements can sometimes
be significant, particularly where large glass windows are involved. Chains of vertical
electrolevel or MEMS beams can be set between reference points to provide x and
y displacements as shown in Figure 7.33, if high resolution is necessary or dynamic
responses to wind and earthquake are required, a high rate of reading will be required,
preferably only triggered by an event. It must be remembered, however, that all buildings respond daily to the sun's traverse and perform a semi-cyclic response which
varies in magnitude with the time of the year. Very sensitive instruments will record
these movements which at times are surprisingly large. The roof of the Mansion House
moved sideways 10 mm on one summer's day.

If less resolution is acceptable, total station surveying, laser scanning and advanced
multi-picture photogrammetry can also be used to determine the movement of points
on the exterior of a building by comparing any two measurement sets. This technique
is minimally invasive and can be used for long-term changes. Care must be taken
to ensure compatible conditions are common to each block of data, as all structures
(as noted above) show daily and yearly movements, which are often of significant
magnitude compared to the settlement changes being monitored.

Finally, in steel structures, the relative column loadings may be significantly changed
by settlement if the structure is very stiff. Strain gauges set on the four faces of a column
can be averaged to provide the load changes during any adjacent work. If the columns
are tubular then three set at 120° in plan are ideal.

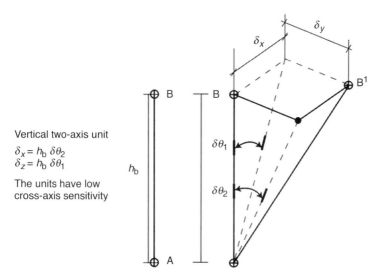

Figure 7.33 Vertical electrolevel/MEMS beam.

Data presentation

The above blocks of data for a critical building will be read at various frequencies as discussed in the introduction according to the relevant situation a), b) or c) above and catagorized. The database of raw information will already be digitised; this can be transmitted to any relevant data store or Internet database and accessed via a networked computer. Raw data are processed, corrected where necessary for temperature sensitivity and the relevant engineering output calculated. Normally, data are presented as the change in location or change in load since the start of the instrument recording, or as a change from any user-set time or any physical value.

Screen presentation of this physical data can be assembled in numerous ways, clearly depending on the sensitivity of the problem, the distortion of the building is usually the key evidence. The typical hierarchy of data is therefore likely to be as follows: Figure 7.34 which shows the zero raw data and raw data at time t_n; Figure 7.35 which shows the actual change in levels at t_n; finally Figure 7.36 which shows the angles of distortion calculated by the conventional method from $3i$ into the trough. The real requirement is the curvature at the point location 6 (Figure 7.37), if we assume an idealised circle ($R =$ constant) as in Figure 7.38, at point X.

The angle we require is small and is the radius at X, i.e. as A–B is parallel to the tangent at X, the angle P–A–B is also β_p, i.e. $\frac{d}{2l}$. Distortion is $2\beta_p$ assuming a uniform radius over length $2l$ see Figure 7.38. In Figure 7.35 the curvature varies from points 4 to 8. But 1, 2, 3, 4 are very nearly a straight line. The conventional system gives a varying angle of distortion from 4 to 8, but commonsense indicates the critical point is between locations 6 and 7.

The critical use in situations arising from Figure 7.36 is to identify the worst local angular distortion and then to estimate the future trends (Figures 7.39 and 7.40).

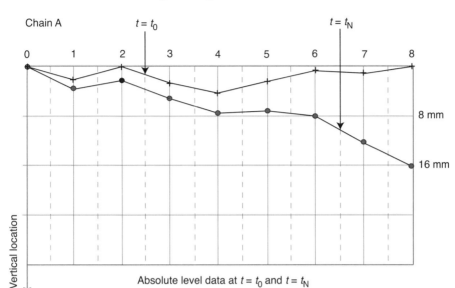

Absolute level – location for $t = t_0$ and $t = t_N$

Figure 7.34 Idealised output for building data, level profile (raw data).

With the data available location 6 appears to be the key location. Figure 7.39 shows a sketch of the angular distortion up to time $t = t_N$ extended in sketch form in various possible ways to A, B or C.

The final possible plot which can stretch the credibility of even the best data is the rate of change of angular distortion with time (Figure 7.40). For the case sketched C and A look unlikely, a change in the B range is probable. The question though is will B exceed the 1.3×10^{-3} critical value? The next four time steps are vital. Obtaining readings to this level for analysis requires extremely high levels of displacement accuracy and very high stability in instruments. In the proposed instrumentation scenarios b) and c) such sophistication may be critical. Similar graphical presentations are possible for all linear data chains.

Multiple data chains such as an updated version of the electrolevel beam system employed at the Mansion House enable the complete basement settlement field to be plotted as a contour plan. Figure 7.31 showed the final settlement on completion of tunnelling. It should be noted that the change in contour spacing clearly suggests a band of angular distortion, but that it is not orientated to the line of any instrument chain. A section at right angles to this band would identify the distortion and its change with time.

When considering a type c) scenario where compensation grouting is involved, Figures 7.41 and 7.42 show an idealised schematic arrangement for arrays of packer grout pipes from a working shaft, the location of in-place horizontal profile gauges between the grout pipes and the building, and between the grout pipes and the tunnel. A matrix of electrolevel (or MEMS accelerometers) beams across the floor of

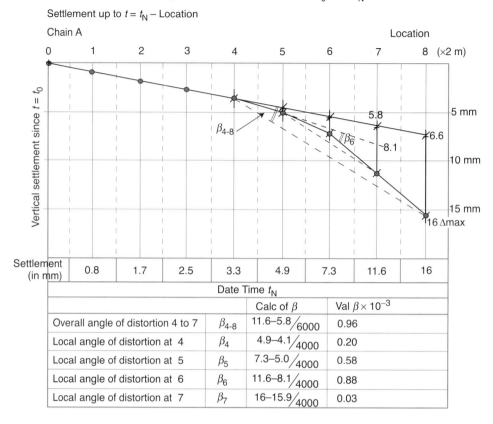

Change in level with location from $t = t_0$ to $t = t_N$

Settlement up to $t = t_N$ – Location

Settlement (in mm)	0.8	1.7	2.5	3.3	4.9	7.3	11.6	16

Date Time t_N		Calc of β	Val $\beta \times 10^{-3}$
Overall angle of distortion 4 to 7	$\beta_{4\text{-}8}$	$\dfrac{11.6-5.8}{6000}$	0.96
Local angle of distortion at 4	β_4	$\dfrac{4.9-4.1}{4000}$	0.20
Local angle of distortion at 5	β_5	$\dfrac{7.3-5.0}{4000}$	0.58
Local angle of distortion at 6	β_6	$\dfrac{11.6-8.1}{4000}$	0.88
Local angle of distortion at 7	β_7	$\dfrac{16-15.9}{4000}$	0.03

Figure 7.35 Settlement profile at t_N.

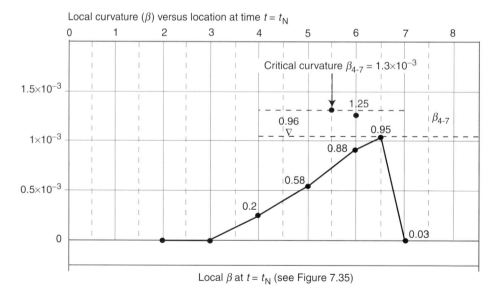

Figure 7.36 Curvature profile (angle of distortion) at t_N.

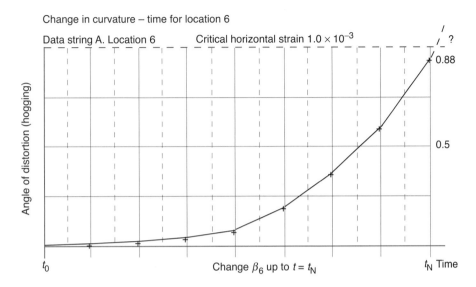

Figure 7.37 Change in angle of distortion at location 6 up to time t_N.

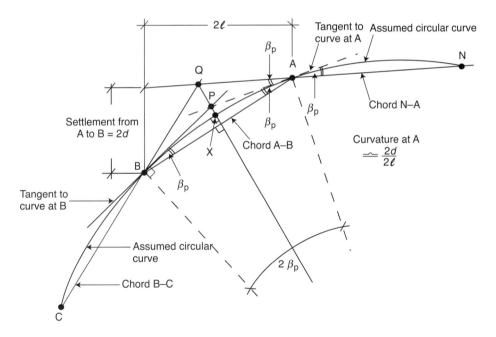

Figure 7.38 Idealised angle of distortion (curvature).

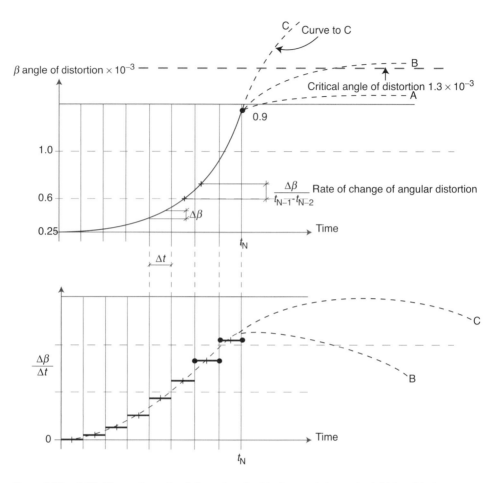

Figure 7.39 and 40 Change in angle of distortion β with time and change in $\Delta\beta/\Delta t$ with time.

the building cover both column location and the floor area (it is very easy to increase a floor area and still fail to compensate a column base).

In this case, the instrumentation data must be processed in near real time and shown on a screen both as an absolute level/change/location plot, and as a rate of level change with time against location plot. These plots can assist in the grouting pressure/packer location control, and the instrumentation engineer must make decisions, again in real time. Before tunnelling and compensation grouting, the whole system must be stabilised and datumed. Considerable care must be taken when grouting within the near vicinity of the face, as the grout pressure may inflict undesired stress on the newly formed tunnel crown, face and lining.

In both situations b) and c) any serious problem developing within a structure is almost inevitably triggered within the tunnel itself. If potential collapse is developing in the soil, the tunnel shape will not be deforming in an acceptable manner, and there may be sufficient time to observe and assess the development of unexpected

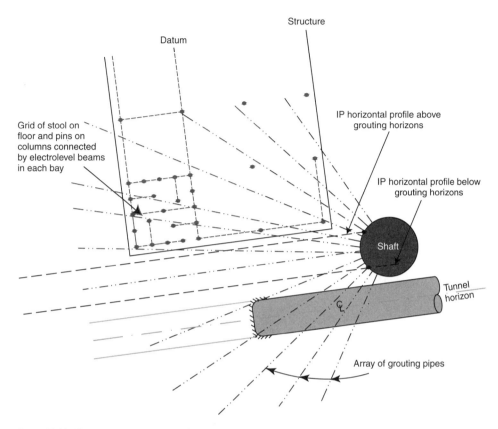

Figure 7.41 Compensation grouting layout.

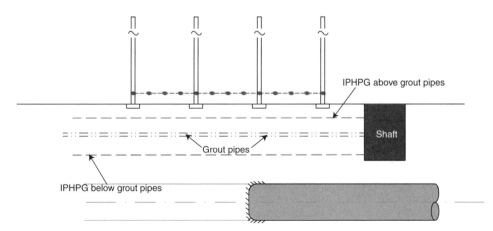

Figure 7.42 Vertical section through the instrumentation array.

convergence distortions, in particular any hint of lining cracking internally or loss of convex curvature. Lining showing a flattened shape will be prone to buckling and this can be very sudden and unexpected.

In such situations, personnel should never be allowed inside the tunnel to carry out measurements or remedial measures. The tunnel lining and subsequently the structure will give very little prior warning of collapse. In a recent well-documented collapse near London, the surface structures above the tunnel showed no significant movement and collapsed totally between the five-minute measurement intervals.

Chapter 8

Data logging, recovery and presentation

Basics

Data from the various instruments available to the industry still come in two formats: direct analogue signals (e.g. millivolt, volt, frequency outputs) or as a digitised string. The flexibility of handling digital strings means that the vast majority of analogue data are now being digitised, either within the instrument or at the collection point. A data component (i.e. a reading) will usually be preceded by a code block which contains some or all of the following:

a) Project and site identifier;
b) date and time of reading;
c) ambient temperature; and
d) ambient pressure.

Hand-booking of data is now limited to remote, rarely read units such as water-level indicators (dipmeters). There are a number of small specialist, handheld, pocket-sized units which directly record and store the specific data type at the press of a button, an example for a vibrating wire recorder is shown in Figure 8.1.

These handheld devices are particularly suitable for readings at single locations, such as multiple anchor rod extensometers, piezometers, electrolevel tiltmeters, etc.

The data stored in the readout/logger is transferred (normally using a download program running on a PC) to the relevant database at the site office. The data are what are termed 'raw data', simply consisting of a number output relevant to the instrument type and the parameter being measured. The site database will use specific programs, each customised (for the interpretation of data from one instrument type such as inclinometers) to reduce the raw data to an engineering value using relevant instrument calibration factors. These data are subsequently compared to the original zero values.

This processed data can be manipulated in many different ways; for example, compared against time, to adjacent instruments, to construction activity, against weather conditions, etc.

The data can be further manipulated for individual instruments to provide rate of change with time. This approach, as indicated earlier, is for the investigation of small differences and requires high quality, stability and resolution in the instrument.

Figure 8.1 A vibrating wire readout/recorder.

The processed rate data can be a key indicator of the approach of incipient failure or indeed (in the other direction) towards stability.

A development is to include analogue-to-digital conversion within each instrument or group of instruments and to send the digital data, which is immune to interference or the effects of cable resistance or capacitance, to a datalogger or ruggidised PC (for example a Campbell CR1000). When installed, a field logger such as this can operate for at least a month on a set of batteries, or almost indefinitely if connected to a solar panel.

A technician can collect data whenever required or via a suitable peripheral device connected to the datalogger; data can be collected via GSM modem, landline, direct radio link, local WiFi network or satellite up/downlink.

The data are then manipulated, as discussed earlier, within the authorised user's office to provide appropriate engineering unit data, which can subsequently be made available to other appropriate groups as graphical information on a PC or as part of an instrumentation and monitoring report.

The key component other than the individual instruments and the datalogger is the wiring system for:

a) Instrument power supply;
b) the transmission of analogue signals or digital data to the datalogger; and
c) any interface (such as a terminal box) between the instruments and the logger.

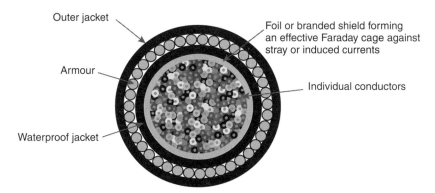

Figure 8.2 Section through a typical high-quality instrumentation cable.

Civil engineering sites are wet, dirty and dusty, often have extremes of temperature, and are in a constant state of flux. Cable routes and datalogger locations are often difficult to define and locate, and to keep intact once installed. One of the most common problems facing instrumentation installers is damage to these components by other site personnel. Careful planning and protection are essential and specific to each individual site, and these need to be carefully documented and all parties made aware of the vulnerability of the component parts.

A typical instrument cable section is shown in Figure 8.2.

Cables can be of considerable length and this can cause degradation of some analogue signals (vibrating-wire signals are relatively immune), this is usually not a problem with digital data. As a general guide, the maximum recommended cable lengths for each instrument type are:

- Analogue instrumentation readings, e.g. electrolevel or potentiometer −250 m;
- vibrating-wire instrument readings – up to 5 km; and
- digital data – dependent on format used (e.g. RS-232, 485 etc.) – up to 5 km.

Radio connection

Between 2003 and 2008, the release of unlicensed low-power radio frequencies such as 433 MHz, 869 MHz and 2.4 GHz for data transmission has enabled the replacement of cables in certain areas. Small, secure transceiver units have been developed which can be mounted adjacent to the instrument or at the top of an instrumentation borehole (e.g. an IPI chain) and these can receive instructions to start logging the instrument and then transmitting data back to a master station consisting of a receiver and a datalogger.

Two types of radio systems are now available, known as a star network and a mesh network. Star network systems (Figures 8.3 and 8.4) typically comprise transmitters which send data at pre-determined intervals to a centralised receiving station. These systems can be very robust but have limitations, in that, once transmitted, if data are not received they are lost, and time synchronisation can be poor between individual sample instruments.

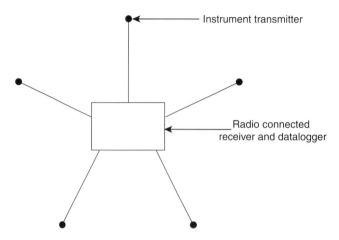

Figure 8.3 Typical star network.

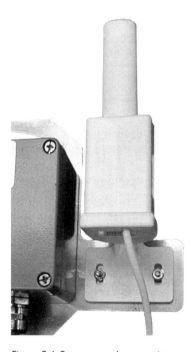

Figure 8.4 Star network transmitter.

The star network transmitters typically use the 433 MHz or 869 MHz band and require line of sight between the transmitters and the receiver, although some reflection and refraction from structures can be accommodated. The maximum range between transmitter and receiver is highly dependent on site conditions, but in practice between 50 and 250 m has been observed.

Mesh networks

Mesh networks (Figure 8.5) are the latest development and far more complex, they have the ability to share data around the network, not simply transmit it in a 'scatter-gun' approach to a receiver. As each element (commonly known as a node or module) is both a receiver and transmitter (i.e. a transceiver) time synchronisation is very precise, data are very rarely lost (if at all) and many logging parameters can be changed around individual parts of the system – for example, having differing read/store rates for different instrument groups within the system.

As each mesh network module (Figure 8.6) is a transceiver, as long as one or more modules can 'see' an adjacent module, the limitations of the star network are largely removed. Mesh networks operate at the higher 2.4 GHz frequency and a typical point-to-point (i.e. module-to-module) range is 50–125 m. Again, as long as two or more modules can communicate, line of sight is not required between all the modules and the receiver (which in a mesh network is termed a Gateway), and on to the datalogging facility. Again, data can be routed to authorised users by various mobile, landline or satellite links, or via the Internet.

When considering such a system, care must be taken to understand the exact nature of the radio system under consideration, and in particular whether the transmitted reading is analogue or digital. Also does the analogue to digital converter within the device have sufficient resolution (i.e. bits) for use in geotechnical and structural monitoring? Typically, 16 bits should be considered a minimum.

WiFi and IP address sensors

Mesh networks are a great advance; for the next generation of sensors we are looking to use WiFi for communication. WiFi is a global standard which accepts information

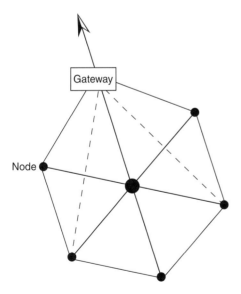

Figure 8.5 Typical mesh network.

Figure 8.6 Mesh network module.

from computers, mobile devices, dataloggers, etc., and allows access to networks and the Internet. Originally confined to small areas and within specific buildings, such as offices, it could accept data and provide the required output. The parallel growth of mobile devices requiring ever-increasing and large quantities of real-time data (such as video streaming) has led to WiFi becoming the world standard of communication for almost all data streams.

As WiFi local area networks expand, costs are falling. Cities such as Taipei in Taiwan are already fully networked for public use, enabling WiFi to accept individual sensor data, each sensor having an individual IP address.

Internet-based software and management systems are becoming more prevalent and are able to handle large volumes of data with maximum ease, reliability and speed.

Internet-based data presentation and underground construction information management systems (UCIMS)

Instrumentation has become automated, with complex front-end software used to process and display the large volumes of data that have become common. The first generation of stand-alone software applications resided on PCs, typically located in the site office. Whilst this was an advance on a spreadsheet, sharing data between engineers could be difficult, and remote access especially so. Dedicated networks and specific lines of communication had to be set up.

The development of the Internet and programming languages made it possible to upload raw data from dataloggers to a site-based computer which processed the data and displayed it on a website, now the web server can carry out the processing and develop the display of data. Automated alarms and reports (typically in pdf format) can also be generated. This has greatly reduced the repetitive component of a monitoring engineer's workload and has allowed him or her to identify and concentrate on areas of concern.

The development of systematic onscreen coding, such as single coloured boxes for each instrument type together with alert codes (conventionally green, blue, amber and red[1]) enable any area of concern to be rapidly identified and, with multilevel passwords, any individual connected to the Internet can view the data.

It is by no means unusual for every individual sensor or groups of sensors on a project to have different alarm levels and on some projects the blue level is omitted.

A typical screenshot from such a system is shown in Figure 8.7.

A further development of this process is to incorporate all the instrumentation software into an integrated management system that contains all monitoring data for an ongoing project, such as manual survey, tunnel-boring machine parameters, photographic records, etc. These systems are termed 'underground construction information management systems' (UCIMS) and need extremely flexible indexing, cross-referencing and retrieval systems, which most importantly also need to be very user friendly.

Very-high-end dedicated web servers are required to handle the volume of data held in these systems, and the simpler graphical style of presentations, shown earlier, have to be interlinked to enable easy navigation through to the interrogator's current point of interest. The system is continually kept up to date as every form of input data is now automatically incorporated as it is obtained.

Access by the hundreds of individuals, engineers, construction managers, clients, consultants and even building owners cannot be unrestricted. Selective access control

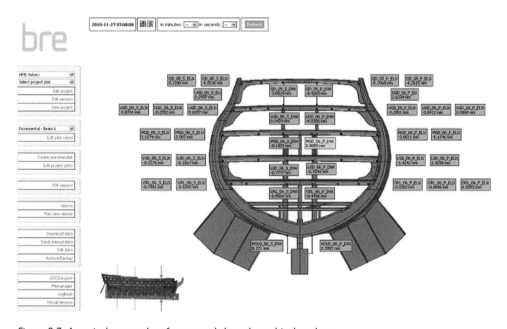

Figure 8.7 A typical screenshot from a web-based graphical package.

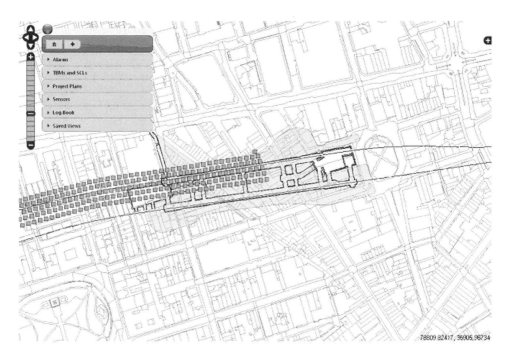

Figure 8.8 A typical screenshot from a GIS UCIMS package.

has to be installed, which gives limited individual permissions, specific to the individual's responsibility, and decision-making and control functions. This is known as permission group management (PGM) and the multi-zoned, multi-level GIS (global information system) technique controls individuals' local search areas and the levels of data they can access. The system should provide the group best able to control and manage any specific risk with every piece of available data in the most direct and easily accessible form, while avoiding the potential for data overload. Parallel to this centralised, highly secure system and database a compatible local area network (i.e. PC-based) system is often used, which enables the individual engineer to download and manipulate data for detailed local assessment.

A typical screenshot from such a web-based system is shown in Figure 8.8.

Chapter 9

Conclusions

The author hopes the reader has found the chapters within this guide informative and useful. The general philosophy of field instrumentation and the conceptual background of most instrument types has been outlined, but the detailed listing of individual products with their ranges and accuracies has not been attempted, as these represent individual companies' commercial data, and can be obtained from the companies' brochures. When planning field instrumentation it is essential to carry out an initial 'class A' prediction, using a parametric approach to soil and structural properties, to obtain estimates of the data ranges and accuracy that each installed instrument will require, and then use the vast range of manufacturers' data sheets to identify and specify suitable units.

Unfortunately, it is common practice for instrument specifications to be copied, without critical assessment, from previous sources; the result can be completely inappropriate criteria for the new work. The key criteria that are always appropriate are stability, reliability and repeatability. Range sensitivity and reading (logging) rate should be site- and location-specific.

An instrument datum can be more difficult to determine with some instruments. Most displacement systems can be linked to a stable reference; loads and stresses are more difficult as the *in situ* state will not be zero stress, so unless the units are factory zeroed, changes are only relative and not related to absolute values.

The author has outlined some examples of instrument layouts and methods of interpreting data. Interaction situations involving the observational method or safety monitoring have been discussed; in these situations the key extra information needing specification is the maximum reading rate and the data resolution level.

However, a realistic value should be taken for the resolution level, it is not very easy when acquiring rate of change data to seek resolution levels that are currently unobtainable.

The most important component (and often the most costly one in any field instrumentation system) is the personnel involved, during installation, monitoring construction, assessing data, reporting and back-analysis. They should understand why the instruments they are setting up and the data they are collecting are relevant; in particular they must be briefed about anomalies they should be aware of.

During the installation in particular, the resulting interpretations will only be as good as the site investigation information allows. Additional local site investigation data can be obtained during installation, but if it is thought beneficial it must be planned, timed, specified and priced, not just left to the often heard phrase 'Oh, wouldn't it be nice if we just ...'

Finally, all projects should be assessed and written up, and published on completion (i.e. not filed and lost) so that the industry, or at least those involved, may benefit from experience, and future work can therefore be made more relevant, economic and safe.

To summarise:

a) There has been little change in the fundamentals of field instrumentation and surveying over the last 20 years. However, procedures using automatic levelling, total stations, laser technology and photogrammetry have become less expensive and hence their use has increased.

b) Inclinometers, horizontal profile gauges, electrolevel or accelerometer beam systems have also vastly benefited from automation and increased reliability. Their stability and accuracy has also improved to a level where rate-dependent data can be assessed.

c) The development of wireless systems and WiFi standards has speeded up installation, and removed the mass of vulnerable wiring.

d) Water-level systems and pore pressure measurement remain much as they have been, and their data are simple, reliable and accurate.

e) The use of fibre-optics is new and exciting.

f) The vast storage and rapid processing ability of modern PCs, together with advanced graphics, is transforming the presentation of data, but in many ways is making the engineer's decisions even more stressful.

g) The industry demand is for cheaper, more rapid construction. By implication, risk levels must increase. Assessing and monitoring risk requires instrumentation which must match the resolution required by the decision-making process, and be sensitive to, for instance, the tunnel/lining/soil/structure interaction relationships.

h) We need to continue, not just to improve instrumentation, but the availability, cost and ease of setting up three-dimensional computer simulations. We need to increase the influence of parametric data changes, develop believable, advanced, constitutive models for soils, and have a site investigation that can give us input parameters.

i) Until back-analysis confirms the above points, the adoption of the 'observational method' will remain the practical engineer's response to accounting pressures. The engineer deserves the rewards not the accountant!

Notes

Chapter 3

1 Fibre-optic sensors can now measure continuous strains along individual fibre optic lines (see Chapter 6, page 144).
2 The local barometric pressure can also influence the water level and an air line connection between the datum house and the reading unit is effectively incorporated via the pipe B.

Chapter 4

1 The frequency is excited by passing an AC current, possibly of square wave form, through the coil in a swept range from below the expected fundamental frequency of the wire to some 50–60% above. It must not rise to twice the fundamental frequency as this would result in inducing the first harmonic frequency and lead to errors in the resulting data.

In practice the frequency sweeps lie between 400Hz and 6kHz, usually in four bands (to prevent the harmonic problems described above) and are produced as a 5 or 12 V square wave. (The old technique was a single 12 V DC pluck.) Despite the square wave stimulus, the wire's resultant fundamental frequency induced in the coil is a sine wave.

Chapter 8

1 Typical alarm levels are defined as:

- Green: data within expected and acceptable limits.
- Blue: data approaching (50–60%) of the upper limit, keep under review.
- Amber: data approaching (70–80%) of the upper limit, requiring both more regular review, the assessment of risk and the preparation of mitigation measures (modification of the works), or the implementation of an emergency preparedness plan (EPP).
- Red: immediate action required to assess rates of change and implement modification to procedures (including the potential to stop work).

References

Bassett R.H., Kimmance J.P. and Rasmussen C. "An Automated Electrolevel Deformation Monitoring System for Tunnels" *Proceedings Institution of Civil Engineers. Geotechnical Engineering*, 137, pp. 117–25, July 1999.

Bassett R.H. and Rasmussen C. "An Automatic Deformation Monitoring System for Combined *x* and *z* Displacements in Structures and Tunnels." *Proceedings 4th International Symposium Field Measurements in Geomechanics*, Bergamo, Italy, pp. 471–88, 1995.

Boscardin, M.D. and Cording, E.J. "Building Response to Excavation Induced by Settlement" *ASCE Journal of Geotechnical Engineering*, Volume 115, No. 1, January 1989.

Card C.B and Carder D.R "A Literature Review of the Geotehnical Aspects of the Design of Integral Bridge Abutments" TRL Report No.PR52. ISBN 0968-4093.

England, G., Tsang, N. and Bush, D. "Integral Bridges: A fundamental approach to the time temperature loading problem" February 2000. ISBN 978-0-7277-3541-6.

Forbes J., Bassett R.H. and Latham M.S. "Monitoring and Interpretation of Movement of the Mansion House due to Tunnelling" *Proceedings Institution Civil Engineers. Geotechnical Engineering*, 107, pp. 89–98, April 1994.

Harris D.I., Mair R.J., Love J.P., Taylor R.N. and Henderson T.O. "Observations of Ground and Structure Movements for Compensation Grouting during Tunnel Construction at Waterloo Station" *Geotechnique*, 44, pp. 691–713, 1994.

Kimmance J.P. "The Development and Testing of a Prototype Automated Electrolevel Deformation Monitoring System" *Jubilee Line Extension*, Internal Report, 104-DCN-GEN-RP-074, 30 pages, 1995.

Price G., Wardle I.F., Frank R. and Jezequel J.F. "Monitoring the Below Ground Performance of Laterally Loaded Piles" *Ground Engineering*, pp. 11–15, July 1987.

Price G. and Wardle I.F. "Comparison of Vertical and Shallow Angled Raked Piles under Lateral Load" *Transport and Road Research Laboratory, Contractors Report* 45, p. 78, 1987.

Price G., Longworth T.I. and Sullivan P.J.E. "Installation and Performance of Monitoring Systems at the Mansion House" *Proceedings Institution Civil Engineers. Geotechnical Engineering*, 107, pp. 77–87, April 1994.

Price, G. and Wardle, I.F. "The Use of Computer Controlled Monitoring Systems to Control Compensation and Fracture Grouting" *Institution Civil Engineers, Proceedings Conference Grouting in the Ground*, Thomas Telford, London, pp. 203–14, 1992.

Rasmussen, C. "Electrolevel Test Data" Presented to the discussing session of Instrumentation in Geotechnical Engineering, Hong Kong, May 1995. Available from itmsoil Ltd SINCO, Arley Tunnel. *The Indicator Newsletter*, No. 1, pp. 1–2, 1994.

Rasmussen, C.P., Wood, R. and Wong, E. "Characterising the Performance of Electrolevel Sensors" Hong Kong I.C.E. Instrumentation Conference, May 1995.

Mao, J. and Nindl, D. "White Paper – Surveying Prisms, Characteristics and Influences" Leica Geosystems, March 2009.

Index